SAI SPEED MATH ACAI

ABACUS MIND MATH

Step by Step Level – 2 Guide to Excel at Mind Math with Soroban, a Japanese Abacus

LEVEL – 2

INSTRUCTION BOOK

FOR WORKBOOKS 1 AND 2

PUBLISHED BY SAI SPEED MATH ACADEMY

USA

www.abacus-math.com

Copyright ©2014 SAI Speed Math Academy

All rights reserved. No part of this publication may be reproduced, transmitted, scanned, distributed, copied or stored in any form or by any means, electronic, mechanical, photocopying, recording, or otherwise, without prior written permission from SAI Speed Math Academy. Please do not participate in or encourage piracy of copyrighted materials which is in violation of the author's rights. Purchase only authorized editions.

Except in the United States of America, this book is sold subject to the condition that it shall not, by way of trade or otherwise, be lent, re-sold, hired out, or otherwise circulated without SAI Speed Math Academy's prior consent in any form of binding or cover other than that in which it is published and without a similar condition, including this condition being imposed on the subsequent purchaser.

Published in the United States of America by SAI Speed Math Academy, 2014

The Library of Congress has cataloged this book under this catalog number:
Library of Congress Control Number: 2014907005

ISBN of this edition: 978-1-941589-03-8

Thanks to **Abiraaman Amarnath** for his valuable contribution towards the development of this book.

www.abacus-math.com
Edited by: WordPlay
www.wordplaynow.com
Front Cover Image: © [Yael Weiss] / Dollar Photo Club
Printed in the United States of America

Our Heartfelt Thanks to:

Our

Higher Self,

Family,

Teachers,

And Friends

For the support, guidance and confidence they gave us to…

…become one of the rare people who don't know how to quit. (-Robin Sharma)

KIND REQUEST

We believe knowledge is sacred.

We believe that knowledge has to be shared.

We could have monopolized our knowledge by franchising our work and creating wealth for ourselves. However, we choose to publish books so we can reach more parents and teachers who are interested in empowering their children with mind math at a very affordable cost and with the convenience of teaching at home.

Please help us know that we made the right decision by publishing books.

- We request that you please buy our books first hand to motivate us and show us your support.
- Please do not buy used books.
- We kindly ask you to refrain from copying this book in any form.
- Help us by introducing our books to your family and friends.

We are very grateful and truly believe that we are all connected through these books. We are very grateful to all the parents who have called in or emailed us to show their appreciation and support.

Thank you for trusting us and supporting our work.

With Best Regards,

SAI Speed Math Academy

Dear Parents and Teachers,

Welcome to LEVEL – 2 of ABACUS MIND MATH training. We hope you had a wonderful time learning and mastering the LEVEL – 1 concepts with your students.

Thank you very much for choosing to use this LEVEL – 2 INSTRUCTION BOOK to learn, teach and excel at mind math using the Japanese abacus called the "Soroban". This is our humble effort to bring a systematic instruction manual to help introduce children to soroban. This LEVEL – 2 instruction book deals with addition using 10 exchange concepts also called big friends. Concepts from LEVEL – 1 will also be reinforced.

This book is the product of over six years of intense practice, research, and analysis of soroban. It has been perfected through learning, applying, and teaching the techniques to many students who have progressed and completed our course successfully.

We are extremely grateful to all who have been involved in this extensive process and with the development of this book.

We know that with *effort, commitment* and *tenacity,* everyone can learn to work on soroban and succeed in mind math.

We wish all of you an enriching experience in learning to work on soroban and enjoying mind math excellence!

We are still learning and enjoying every minute of it!

We would like to thank you again for all your support and encouragements.

GOAL AFTER COMPLETION OF LEVEL 2 – WORKBOOKS 1 AND 2

On successful completion of the two workbooks students would be able to:

1. Add any two to three digit numbers that involve carry-over or regrouping problems.

HELPFUL SKILLS

- Must have mastered LEVEL – 1 concepts

PRACTICE WORKBOOKS FOR STUDENTS

There are two workbooks available for students to practice on the concepts given in this LEVEL – 2 Instruction book. Complete Workbook – 1 before proceeding to Workbook – 2. These Workbooks are **sold separately** and are available under the titles:

Abacus Mind Math Level – 2 Workbook 1 of 2 – ISBN: 978-1-941589-04-5

Abacus Mind Math Level – 2 Workbook 2 of 2 – ISBN: 978-1-941589-05-2

WE WOULD LIKE TO HEAR FROM YOU!

Please visit our Facebook page at **https://www.facebook.com/AbacusMindMath**. Contact us through **http://www.abacus-math.com/contactus.php** or email us at **info@abacus-math.com**

We Will Award Your Child a Certificate Upon Course Completion:

Once your child completes the test given at the back of the workbook – 2, please upload pictures of your child with completed test and marks scored on our Facebook page at **https://www.facebook.com/AbacusMindMath**, and at our email address: **info@abacus-math.com**

Provide us your email and we will email you a personalized certificate for your child. Please include your child's name as you would like for it to appear on the certificate.

LEARNING INSTITUTIONS AND HOME SCHOOLS

If you are from any public, charter or private school, and want to provide the opportunity of learning mind math using soroban to your students, please contact us. This book is a good teaching/learning aid for small groups or for one on one class. Books for larger classrooms are set up as 'Class work books' and 'Homework books'. These books will make the teaching and learning process a smooth, successful and empowering experience for teachers and students. We can work with you to provide the best learning experience for your students.

If you are from a home school group, please contact us if you need any help.

Contents

LEVEL 2 – INSTRUCTION BOOK

TOPICS COVERED

In LEVEL – 2 students will be introduced to concepts involving addition formulas utilizing combination facts of 10 (big friends). They will be learning formulas for numbers +9 to +1. The small friend concepts learned in LEVEL – 1 will be extensively used in LEVEL – 2, so please make sure that the children understand all LEVEL – 1 concepts very well before starting them on LEVEL – 2.

This instruction book helps in explaining Level – 2 concepts with example and sample problems. Practice work for this level is sold separately as Level 2 – Workbook 1 and Level 2 – Workbook 2.

Let your students use the Level 2 – Workbook 1 & 2 as you go through the lessons one by one in the order given while teaching them.
*(**Workbook 1** – has more work for the concepts found in Lesson 1 to Lesson 9.*
***Workbook 2** – has more work for the concepts found in Lesson 10 to Lesson 15.*
Practice work for concepts found in Lesson 16 and 17 are provided in this Instruction Book only.)

FINGERING

Correct fingering is very important, so practice moving earth beads and heaven beads using the correct fingers.

INSTRUCTIONS GIVEN WITHIN THE VIOLET BOX

General information to be remembered while working on the abacus is given here.

INSTRUCTIONS GIVEN WITHIN THE PINK BOX

Useful information in the process of teaching can be found in the pink box.
The most likely mistakes that a child tends to make are also explained here with ways to rectify them as needed.

Each and every child is unique in his/her own respect. His/her understanding of any new concept is also going to be vastly different. So, expect children to surprise you with unique questions and mistakes. With persistent practice, all the hurdles can be overcome.

INSTRUCTIONS GIVEN WITHIN THE RED BOX

Suggests and explains skill building activities that will help improve understanding of that particular week's concept.

BEAD COLORS

= Beads that are not involved in the calculation or the game

= Beads that were already in the calculation or the game

= Beads that have just been moved to ADD in the game

= Beads that have just been moved to MINUS from the game

HOW TO EXPLAIN AN EXAMPLE PROBLEM TO STUDENTS

Teachers: Practice the example problems on your abacus.

To Teach: Example 1: +21

1. Call out the number you are going to add or subtract
2. Say what you are doing on the abacus as explained in the 'Action' column of the example. Each example is explained in detail. Judge your child's maturity and skill level and explain according to their needs.

Clear	Problem	Action
Step 1	**+ 21**	Move **2 earth beads up** to touch the beam on the **tens rod**. Move **1 earth bead up** to touch the beam on the **ones rod.**

HOW TO READ THE HAND PICTURE

While introducing addition fact families which total to 10, use your hands to help children understand the relationship between numbers (1+9=10, 2+8=10, 3+7=10, 4+6=10, and 5+5=10).

While introducing Big Friends Combination formulas (example: +9 = +10 – 1) use both of your hands to represent +10 with 10 fingers stretched out *(to represent adding of one earth bead on the tens rod)* and fold fingers when you minus the combination friend.

Example: +9 = +10 – 1

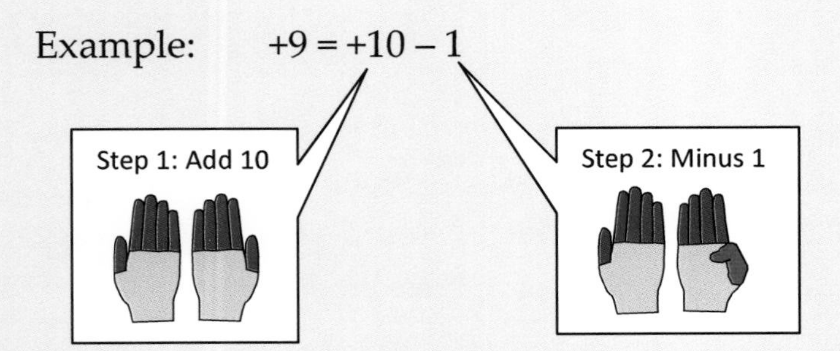

In the above example +10 is a tens place number so add ten on the tens place rod. The second half of the formula tells you to – 1, and one is an ones place number and so – 1 has to be done on the ones place rod.

By using fingers, it is easy for children to memorize the fact family and also understand the big friend combination formulas. When using fingers to represent numbers adding up to 10, each finger is equal to 'one'. When using fingers to represent numbers adding up to 100, each finger is equal to 'ten'. Please do not worry that it would confuse your child because children will easily understand the concept. With practice, they will perfect their understanding.

Alternately, you may choose to use any other items like marbles, counters, beans etc. to teach.

LESSONS 1 – Introduction to big friends combination numbers and facts adding up to 10 and 100.

LESSONS 2 to 15 – Big friend addition formulas for +9, +8, +7, +6, +5, +4, +3, +2, +1, and carryover from ones rod to hundreds rod explained with examples. Each lesson has sample problems to help teachers learn and also introduce your students to a particular week's formula. Use these sample problems to teach your students before allowing them to use their corresponding workbooks.

MIND MATH – Mind math is introduced with simple problems and students who have done well with LEVEL – 1 mind math will be able to do these problems with ease.

If student has trouble with visualizing and computing in his/her mind, try any of the methods as explained in the LEVEL – 1 Instruction book. Do not try mind math on a challenging problem more than 3 times. If any of the methods as explained in LEVEL – 1 do not help, please do not stress students. Let them work on the abacus part of the practice work for a few days and learn the formula well. Once they are comfortable computing using the formula on the abacus, you can try going back to mind math.

Motivate and encourage them by letting them compute with only two set of numbers to begin with. A time will come when they are willing and able to do more complex mind math problems. Maintaining a nurturing attitude towards students is extremely important for success. Patience and tenacity is the key to success with mind math in this level.

EXCLUSIVE BONUS INTERMEDIATE LEVEL LESSONS

LESSONS 16 – Using big friend addition formulas on hundreds place rod is explained with example and sample problems. Practice work for students is given ONLY in this ABACUS MIND MATH LEVEL 2 – INSTRUCTION BOOK. As teachers and parents you will know better about your child and choose to introduce this lesson when they are ready.

LESSONS 17 – Using big friend addition formulas to carryover to the thousands rod from ones place rod or tens place rod is explained with example and sample problems. Practice work for students is given ONLY in this ABACUS MIND MATH LEVEL 2 – INSTRUCTION BOOK. As teachers and parents you will know better about your child and choose to introduce this lesson when they are ready.

**The big friends concepts introduced in this level are like framework for a building, thorough understanding of these facts is vital for their success with multiplication. All the formulas from LEVEL – 1 and LEVEL – 2 will be included in LEVEL – 3. So, please make sure that students understand LEVEL – 2 concepts very well to ensure success with higher levels.*

ATTRIBUTES TO SUCCEED

ATTRIBUTE	DESCRIPTION	PARENTS/TEACHERS	CHILDREN
INTEREST	The state of wanting to know or learn about something or someone	Teaching to instill natural curiosity in children.	Learning to look to their adult teachers for guidance about the things they are curious about.
COMMITMENT	Pledge or bind (a person or an organization) to a certain course or policy	• Setting a time and place for your children every day to practice on their abacus without distractions. • Reading the instructions, and guiding and teaching until the student understands the concept.	• Learning that they are acquiring unique skills that many in their peer group do not possess. • Enthused when they understand that they are able to work out their math problems much faster without the use of calculator.
PATIENCE	The capacity to accept or tolerate delay, trouble, or suffering without getting angry or upset.	Directing children's attention back to work when they get sidetracked by TV or other distractions.	Slowly developing with consistent practice.
TENACITY	Persistent determination	"Winners never quit, quitters never win." • Teaching the value of consistent hard work with positive encouragement.	Developing the habits of persistence and hard work.
GUIDANCE	Supervised care or assistance	Providing instruction and guidance until the student learns to work independently.	Reading and understanding concepts, and ultimately working independently
REWARD	A thing given in recognition of service, effort, or achievement.	• Developing and strengthening the parent child bond by spending time together. • Teaching and enriching your child's life and learning experience.	• Learning a lifelong skill • Improving concentration • Enhancing memory power • Gaining self confidence • Developing positive self esteem

PLACE VALUE OF RODS

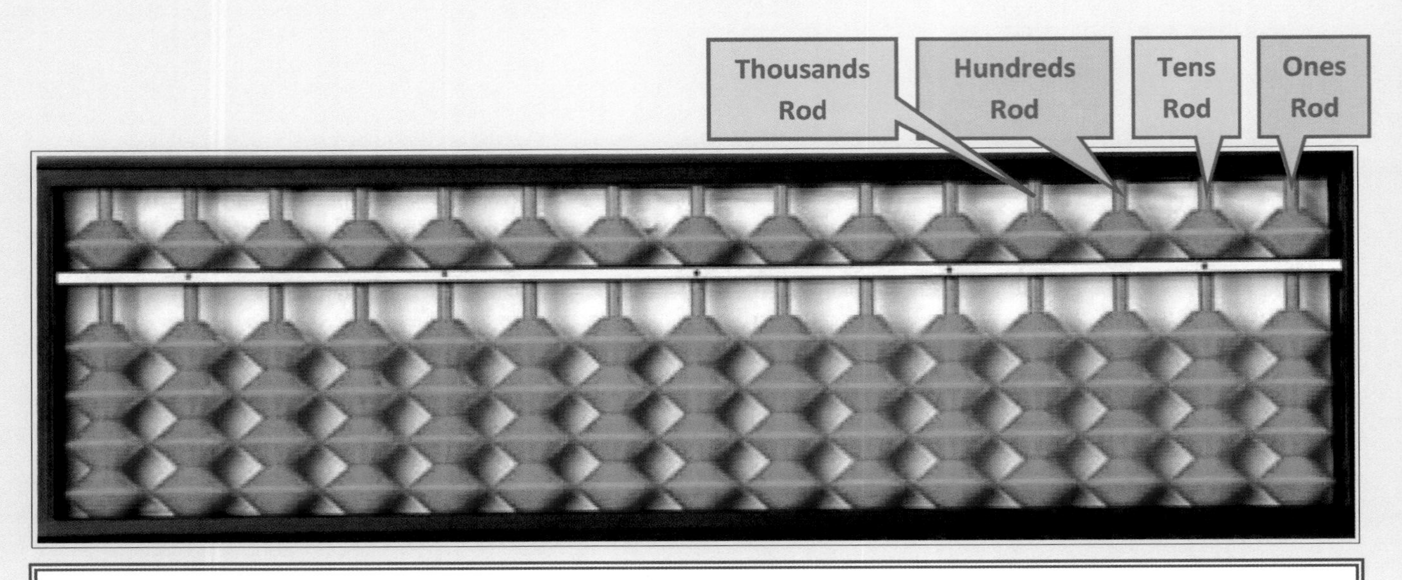

First rod from right is for the **ones place number**. All the ones place number will be set on this rod.

Second rod from right is for the **tens place number**. All the tens place number will be set on this rod.

Third rod from right is for the **hundreds place number.** All the hundreds place number will be set on this rod.

ORDER OF OPERATION

LEFT TO RIGHT: When working with two digit numbers: always add or subtract the tens place number first and then work on the ones place number.

REVISION OF LEVEL 1 CONCEPTS

SMALL FRIEND FORMULAS

TO ADD	TO MINUS
+ 1 = + 5 – 4	**– 1 = – 5 + 4**
+ 2 = + 5 – 3	**– 2 = – 5 + 3**
+ 3 = + 5 – 2	**– 3 = – 5 + 2**
+ 4 = + 5 – 1	**– 4 = – 5 + 1**

WEEK 1 – LESSON 1 – BIG FRIEND COMBINATIONS

POINTS TO REMEMBER

GOAL: For your child to know the **two numbers that add up to ten and hundred**. There are five different combinations for 10 which are 9 & 1, 8 & 2, 7 & 3, 6 & 4, and 5 & 5. There are also five different combinations for 100 which are 90 & 10, 80 & 20, 70 & 30, 60 & 40, and 50 & 50.

Make sure that the students understand the relation between big friend combination numbers. All the formulas introduced in LEVEL – 2 will be using these big friend relationship facts to attain the desired results of adding numbers. Also, combination facts for numbers that are multiple of 10 can be figured out using these big friend combination numbers.
Example: For 100 is 90 & 10, 80 & 20, 70 & 30, 60 & 40, and 50 & 50.
Example: For 1000 is 900 & 100, 800 & 200, 700 & 300, 600 & 400, and 500 & 500.

** For this level, it is enough if students know the combination facts for 10 and 100.*

BIG FRIEND COMBINATION OF 10

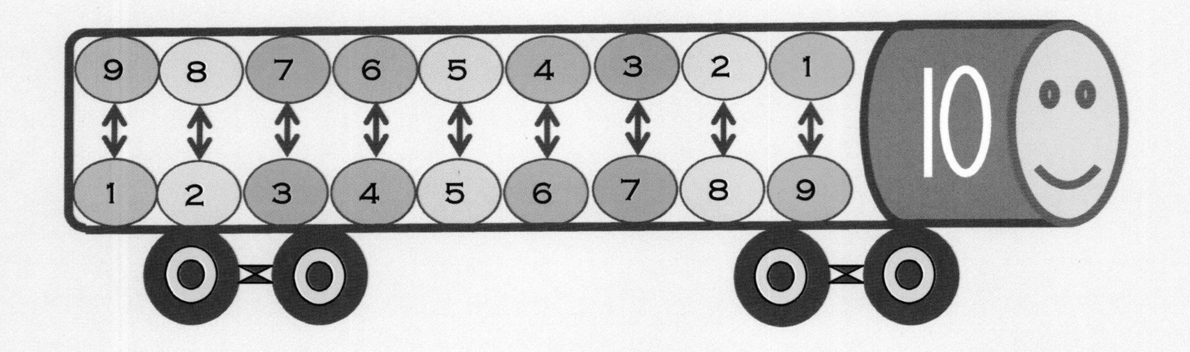

NUMBER SENTENCE FOR COMBINATIONS OF 10
Addition Facts

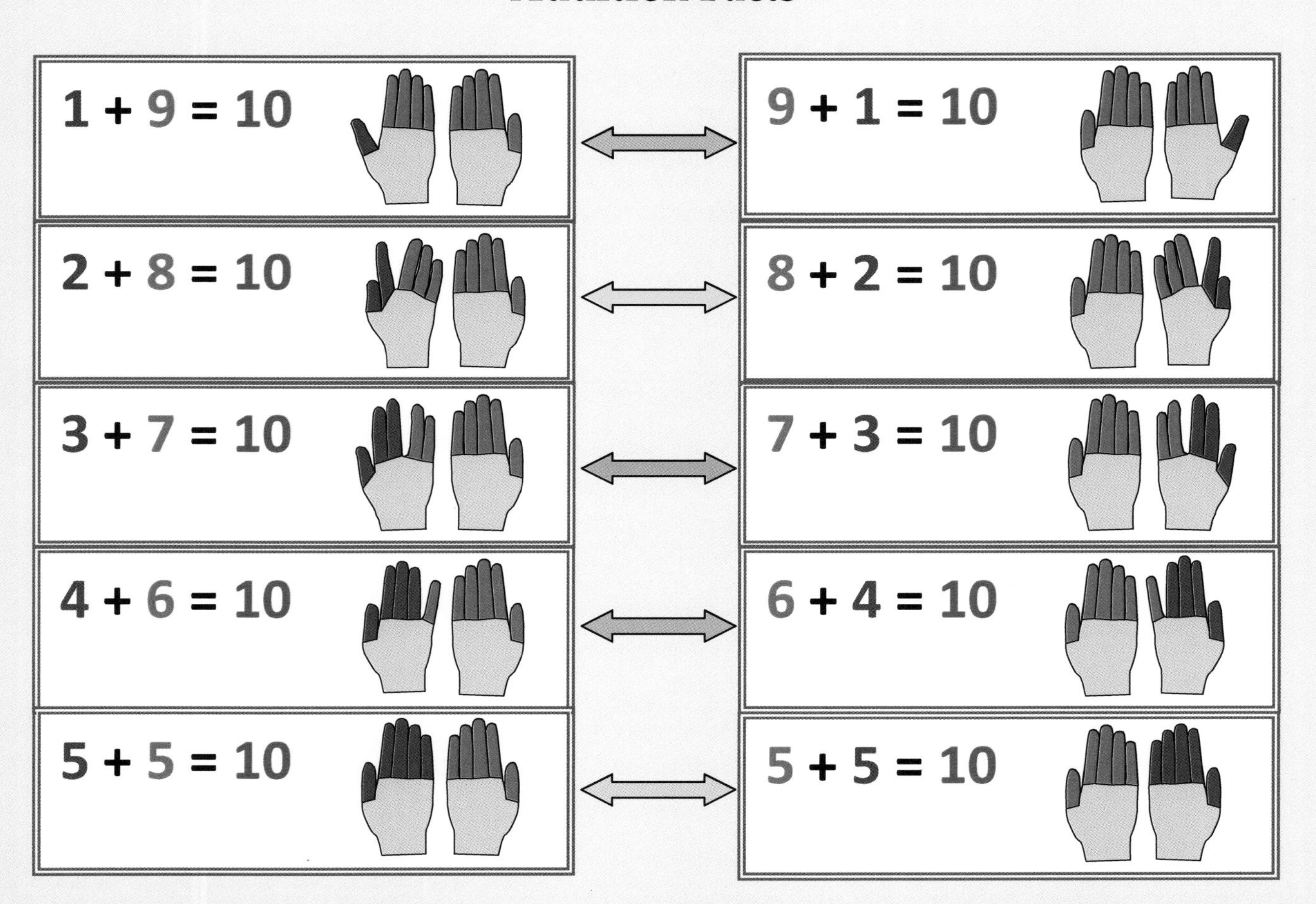

BIG FRIEND COMBINATION OF 100

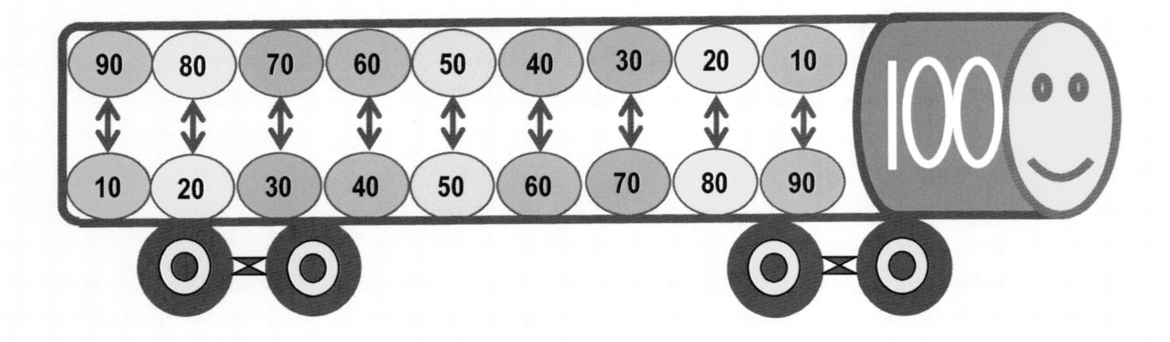

NUMBER SENTENCE FOR COMBINATIONS OF 100

Addition Facts

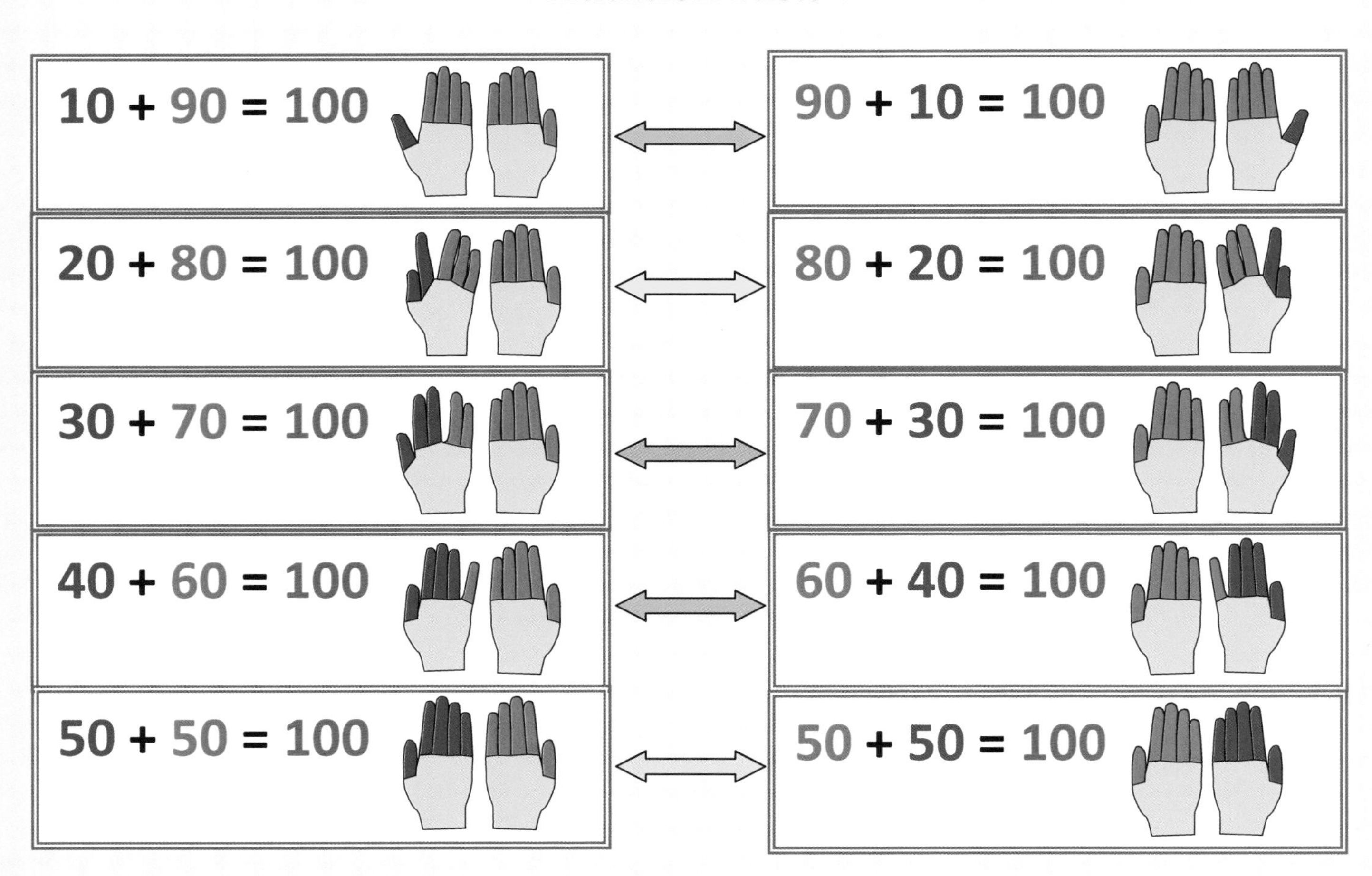

By using fingers, it is easy for children to memorize the fact family and also understand the big friend combination formulas. When using fingers to represent numbers adding up to 10, each finger is equal to 'one'. When using fingers to represent numbers adding up to 100, each finger is equal to 'ten'. Please do not worry that it would confuse your child because children will easily understand the concept. With practice, they will perfect their understanding.

BIG FRIEND FACTS OF 10

1 and 9 are big friends, because together they make 10.
2 and 8 are big friends, because together they make 10.
3 and 7 are big friends, because together they make 10.
4 and 6 are big friends, because together they make 10.
5 and 5 are big friends, because together they make 10.

BIG FRIEND COMBINATION FACTS FOR 100

10 and 90 are big friends of 100, because together they make 100.
20 and 80 are big friends of 100, because together they make 100.
30 and 70 are big friends of 100, because together they make 100.
40 and 60 are big friends of 100, because together they make 100.
50 and 50 are big friends of 100, because together they make 100.

EXTRA PRACTICE

- Quiz your child by asking "Who are big friends of 10?" Answer: 1 & 9, 2 & 8, 3 & 7, 4 & 6, and 5 & 5 are big friends.
- Ask "Why 9 & 1are big friends? Their answer should be something like; "Because 9 & 1 together make 10". Quiz them with similar questions for all the five set of big friend combination numbers.
- Confuse them by asking if 1 & 7 are big friends or if 2 & 6 are big friends. The goal is for your child to know the relationship between the numbers. So, his/her answer has to be "no".

Quiz your child on the combination facts adding up to 100.

POINTS TO REMEMBER WHILE USING A FORMULA

1. **Use the formula only when you do not have enough beads to add or subtract.** When you teach a formula, most kids understand and remember this fact. However, some children will try to do the formula even when they can directly add a number. They will +10 and get confused about why they do not have enough beads to send away the big friend of the number they are computing. So, care has to be taken to make students understand that formulas are to be used **only** when they do not have enough beads to compute an answer.

2. **When formula is introduced ask students to +10** (or +100) **first and then remove the corresponding big friend of the number they are trying to add from the game.**

   a) **You can present the formula in a cohesive order for students to follow.**

      When students have to add a number and they don't have enough beads, they can +10 (or +100). However, they are not supposed to +10 (or +100) so they have to send out the extra beads, which will definitely be the big friend of the number they are trying to add.

Example: 4 + 9

Ones rod has 4 earth beads in the game, and the heaven bead is available for the game but, its value is 5 which is not going to be enough. Now, we know that 9 cannot be added directly.

So, we need to get help from 10 by adding it to the game.

- Add 10 to the game.

Known Fact: 9 + 1 = 10 (9 and 1 are big friends of 10)

We know that 10 is 1 more than 9.

So, to keep only 9 in the game we have to remove 1 from the game.

- Take 1 out of the game.

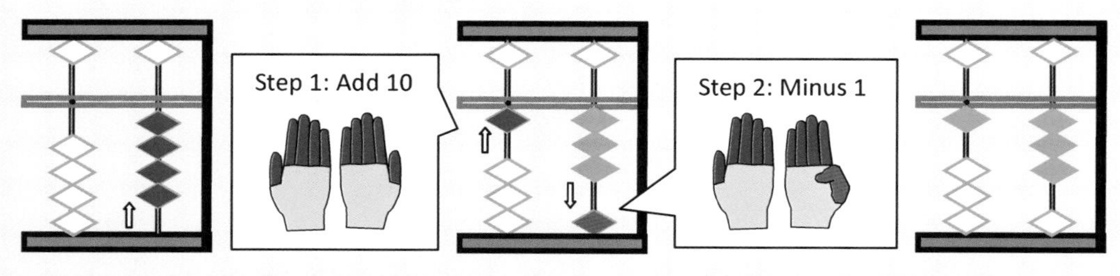

Big friend formulas are easier to remember this way rather than just asking students to remove friend and then add one bead up on the rod to the left – the general way formulas are taught. Students can be taught to follow general method too, but make sure that they understand why they are removing the friend and adding one bead on the rod to the left.

b) Adding 10 or 100 and then removing friend will bring students back to the rod they are working.

When you follow the formula in the order given, students will come back to the same rod they are working on and it is easy for them to move on to the next place value to continue computing.

Example: 22 + 92 = 114

When using the formula to do +90 (9 here is a ten's place number), first +100 and then –10. Now, by following this order their finger is back on the tens place rod and they have to move to the ones place rod to +2.

If students follow the general method they first remove friend on the tens rod (working rod) and then +100, which leaves their finger on the hundreds rod. The next number to compute is +2 for which they have to consciously skip the tens place rod and go to ones place rod. Young students will be challenged in remembering to skip the tens rod and move to ones place rod to finish computing. But like anything, with practice students may overcome this hurdle. Choose the method you think best fits your students' ability to understand and follow the formulas

3. **Visual Aid –** As a visual aid show them both hands with fingers outstretched to represent +10 (or +100), and then fold the fingers to represent minus of the corresponding big friend. By this they can visually see how following the formula helps them add a number indirectly (out of box thinking).

Example: +9 = +10 – 1

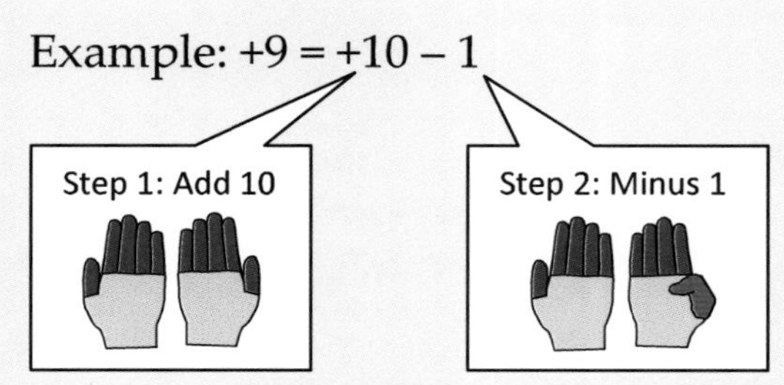

4. **Encourage them to use the reason and figure out the formulas on their own.** After students understand the pattern (s/he will have to be taught a few formulas to get to this point) encourage them to figure out the rest of the big friend formulas on their own. Most students are very eager to guess the next formula and explain the reason as to why the formula works.

5. **This in turn encourages them to stay motivated and increases their self confidence.**

ATTENTION

- Bead is in the game = adding
- Bead is out of the game = subtracting
- Working rod = the rod on which you are trying to add or subtract that corresponds to the place value of the number you are computing.

WEEK 2 – LESSON 2 – INTRODUCING + 9 CONCEPT

1) There are four earth beads on each rod.
2) Value of one earth bead = 1
3) Use your thumb to move the earth bead up (adding) to touch the beam.
4) Use your pointer finger (index finger) to move the earth bead down (subtracting) to touch the frame.
5) There is one heaven bead above the beam (heaven deck) on each rod.
6) Value of heaven bead = 5
7) Use your pointer finger (index finger) to move the heaven bead down (adding) to touch the beam.
8) Use your pointer finger (index finger) to move the heaven bead up (subtracting) to touch the frame.
9) Always set abacus to zero by clearing all the beads away from the beam before starting each calculation.
10) Setting numbers on the abacus:

    Hundreds place numbers go on the hundreds rod.

    Tens place numbers go on the tens rod.

    Ones place numbers go on the ones rod.

LESSON 2 – EXAMPLE

CONCEPTS OF THE WEEK

TO ADD = ADD 10, LESS BIG FRIEND **+ 9 = + 10 – 1** **+ 90 = + 100 – 10**

EXAMPLE: 1

1	After	+ 03	+ 09 (+10 – 1)	= 12
03 09 12	ABACUS LOOKS LIKE	Step 1: Add 10	Step 2: Minus 1	

Problem	Action
+ 03	There is nothing to do on the tens rod because tens place number is zero. Move **three earth beads** to touch the beam on the **ones rod.**
+ 09	There is nothing to do on the tens rod because tens place number is zero. Now we need to +9 on the ones rod, however we do not have enough beads on the ones rod to +9. (*The heaven bead is not in the game, but it does not have nine in it, so we cannot get help from it.*) So, now we need to make use of the fact that **9 + 1 = 10** **When you want to +9 and you do not have enough beads:** **Use +9 = +10 – 1 Big Friend formula to do your calculations.** **Step 1: Add 10** – Move **one earth bead up** to touch the beam on the **tens rod**. *(One earth bead on the tens rod is equal to '10')* *We know that there is a nine in the ten (9 + 1 = 10), so let us get help from 10 by adding it to our game. However, we were supposed to +9, instead we did +10, which means we have 1 more than what we need. So, now we have to send the 1 away from our game.* **Step 2: Minus 1** – Move **one earth beads away** from the beam on the **ones rod**. ***(When we do + 10 and – 1, we get to keep 9 in our game.)*** **+10 – 1 = +9**

EXAMPLE: 2 **+ 90 = + 100 – 10**

1	After + 80 (+100 – 10) + 90 = 170
80 90	ABACUS LOOKS LIKE
170	

Step 1: Add 100

Step 2: Minus 10

Problem	Action
+ 80	Move **three earth beads and the heaven bead** to touch the beam on the **tens rod.** There is nothing to do on the ones rod because ones place number is zero.
+ 90	Now we need to +90 on the tens rod, however we do not have enough earth beads to +90 on the tens rod. **When you want to +90 and you do not have enough beads:** **Use +90 = +100 – 10 Big Friend formula to do your calculations.** *(This can be taught as using the same bead movement as for +9, but add on the hundreds rod and minus on the tens rod.)* **Step 1: Add 100** – Move **one earth bead up** to touch the beam on the **hundreds rod**. **Step 2: Minus 10** – Move **one earth bead away** from the beam on the **tens rod**. There is nothing to do on the ones rod because ones place number is zero.

ATTENTION

- Ask students to say the formula while they use it. This makes it easy for them to understand and follow through with all the steps in the formula.
- When adding 19 to another number (where they need to use +9 formula) students will +10 once and then –1. Make sure they understand that they have to +10 once for the ten in the 19 and **another** +10 to use the formula before finishing with –1 while +9. Ex: 16 + 19
- When following the formula on the tens rod, students usually +100 but then get confused and try to do –1 instead of –10. Make them understand that big friend of 90 is 10 and they help each other.

LESSON 2 – SAMPLE PROBLEMS

TO INTRODUCE +9 = + 10 – 1 FORMULA

Work these problems a few times to study and understand the concept and the relationship between the beads moved.

1	2	3	4	5	6	7	8	9	10
01	02	03	04	12	13	22	21	46	41
09	09	09	19	09	09	19	19	19	19
09	09	09	09	19	09	19	09	01	09

2:S:1

© SAI Speed Math Academy, USA

TO INTRODUCE +90 = + 100 – 10 FORMULA

Work these problems a few times to study and understand the concept and the relationship between the beads moved.

1	2	3	4	5	6	7	8	9	10
14	20	30	40	120	244	130	78	90	240
90	93	90	90	90	90	90	99	99	190
91	93	09	90	90	90	99	99	99	190

2:S:2

© SAI Speed Math Academy, USA

1	2	3	4	5	6	7	8	9	10
55	64	39	44	67	46	88	37	14	135
- 23	94	29	92	31	29	91	99	19	192
99	09	- 46	09	- 64	90	- 59	29	29	99
- 31	19	90	- 41	99	- 23	90	- 11	19	- 214

2:S:3

© SAI Speed Math Academy, USA

POINTS TO REMEMBER

The rows above consist of sample problems to introduce this week's formula. Explain to your child when and how to use the formula. Work with the sample problems until your child understands the formula and that the formulas are to be used ONLY when there are not enough beads to add or subtract.

WEEK 3 – LESSON 3 – COMPLETING + 9 USING SMALL FRIENDS FORMULA

CONCEPTS OF THE WEEK

TO ADD = ADD 10, LESS BIG FRIEND **+ 9 = + 10 – 1** **+ 90 = + 100 – 10**

+9 = +10 –1	**+90 = +100 –10**
GET HELP FROM	**GET HELP FROM**
+10 = +50 –40	**+100 = +500 –400**
–1 = –5 +4	**–10 = –50 +40**

In the previous lesson, all the problems where you had to use the formula were simple and you were able to follow the +9 formula directly. This week, we will see how we might have to get help from our LEVEL – 1 Small Friends formula of +10 and – 1 to do +9. Students will understand this week's concept if you make them understand that they are "getting help" from LEVEL 1 formulas to do +9. You may choose to give a quiz/dictation on lessons nine and ten in LEVEL 1 before introducing them to this week's lesson.

Example 1: 44 + 09 = 53
Set 44 on the abacus. Now, to +9 we need to follow the formula of **+9 = +10 – 1**, but we cannot do +10 directly. So to do +10, you have to follow **+10 = +50 – 40** formula from LEVEL – 1.

Example 2: 35 + 09 = 44
Set 35 on the abacus. Now, to +9 we need to follow the formula of **+9 = +10** –1. Here, you can directly +10 but you do not have enough ones bead on the ones rod to do –1. So to do –1, you get help from the **–1 = –5 + 4** formula from LEVEL – 1.

Example 3: 45 + 09 = 54
Set 45 on the abacus. Here too we need to use the **+9 = +10** – 1 formula. However, both +10 and – 1 cannot be done directly. To do both these steps, you have to get help from LEVEL – 1 Small Friends formulas for **+10 = +50 – 40 and – 1 = –5 + 4.**

Example 4: 350 + 90 = 440
Set 350 on the abacus. To do +90 we now need to use the formula **+90 = +100** –10. Here, to complete the steps you will have to get help from LEVEL – 1 Small Friends formulas for **–10 = –50 +40**.

LESSON 3 – EXAMPLE

EXAMPLE: 1

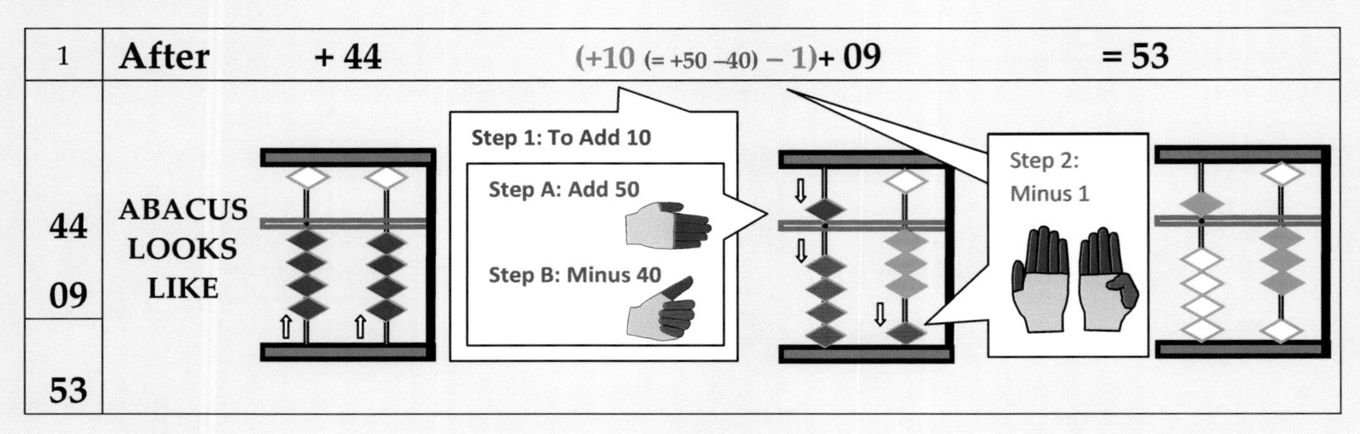

Problem	Action
+ 44	Move all **four earth beads** to touch the beam on the **tens rod.** Move all **four earth beads** to touch the beam on the **ones rod.**
+ 09	There is nothing to do on the tens rod because the tens place number is zero. Now we need to +9 on the ones rod, however we do not have enough beads on the ones rod to +9. *(The heaven bead is not in the game, but it does not have nine in it, so we cannot get help from it.)* So, now we need to make use of the fact that **9 + 1 = 10** **When you want to +9 and you do not have enough beads:** **Use +9 = +10 – 1 Big Friend formula to do your calculations.** ***Now to do +10 you do not have enough earth beads so, GET HELP from small friend formula and do +10.*** **Step 1: Add 10** = **Step A: Add 50** – Move the heaven bead down to touch the beam on the tens rod. **Step B: Minus 40** – Move all four earth beads down to touch the frame on the tens rod. **Now complete the +9 formula by doing – 1 on the ones rod.** **Step 2: Minus 1** – Move **one earth bead away** from the beam on the **ones rod**. ***(When we do + 10 and – 1, we get to keep 9 in our game.)*** **+10 – 1 = +9**

EXAMPLE: 2

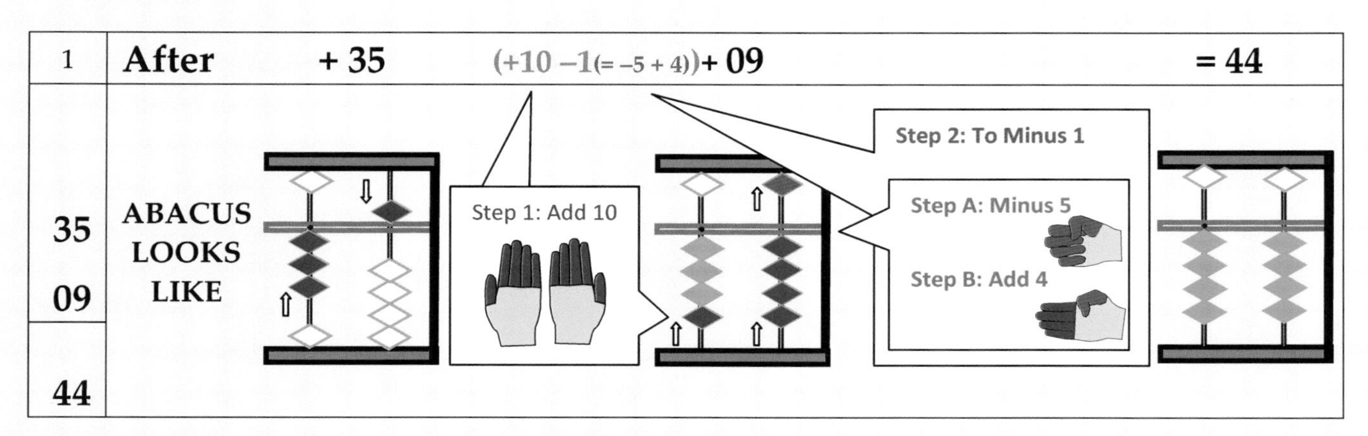

Problem	Action
+ 35	Move **three earth beads** to touch the beam on the **tens rod.** Move the **heaven bead** to touch the beam on the **ones rod.**
+ 09	There is nothing to do on the tens rod because the tens place number is zero. Now we need to +9 on the ones rod, however we do not have enough beads on the ones rod to +9. (*The heaven bead is not in the game, but it does not have nine in it, so we cannot get help from it.)* So, now we need to make use of the fact that **9 + 1 = 10** **When you want to +9 and you do not have enough beads:** **Use +9 = +10 – 1 Big Friend formula to do your calculations.** **Step 1: Add 10** – Move **one earth bead up** to touch the beam on the **tens rod.** *(One earth bead on the tens rod is equal to '10')* ***Now we need to finish the formula by doing – 1 on the ones rod, but we do not have earth beads to do –1 so, GET HELP from small friend formula and do – 1.*** **Step 2: Minus 1** = Step A: Minus 5 – Move the heaven bead up to touch the frame on the ones rod. Step B: Add 4 – Move all four earth beads up to touch the beam on the ones rod. ***(When we do + 10 and – 1, we get to keep 9 in our game.)*** **+10 – 1 = +9**

EXAMPLE: 3

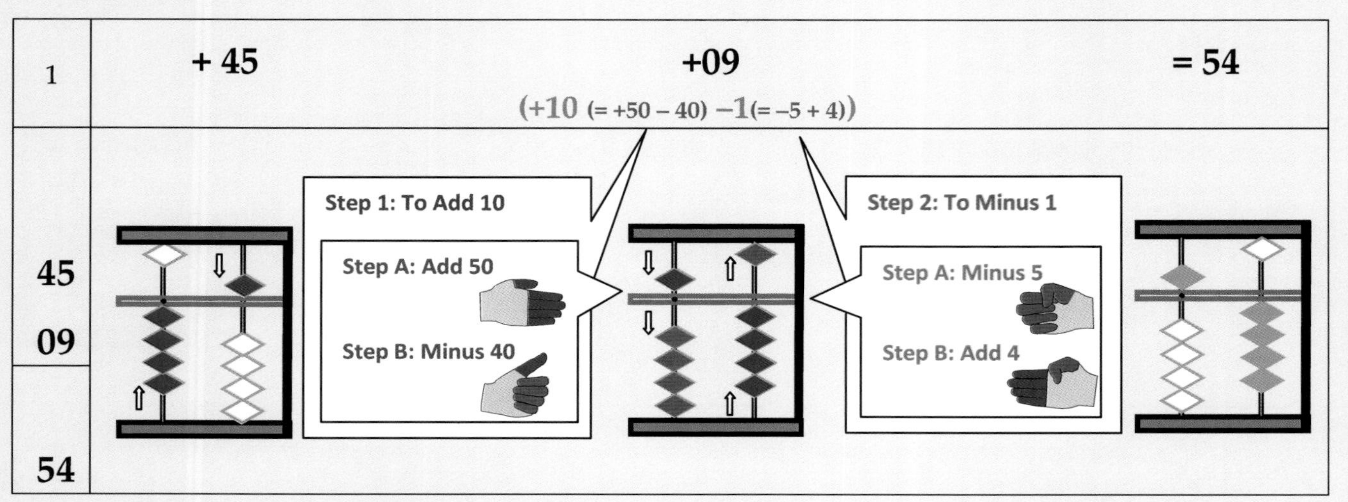

Problem	Action
+ 45	Move **four earth beads** to touch the beam on the **tens rod.** Move the **heaven bead** to touch the beam on the **ones rod.**
+ 09	There is nothing to do on the tens rod because the tens place number is zero. Now we need to +9 on the ones rod, however we do not have enough beads on the ones rod to +9. (*The heaven bead is not in the game, but it does not have nine in it, so we cannot get help from it.*) So, now we need to make use of the fact that **9 + 1 = 10** **When you want to +9 and you do not have enough beads:** **Use +9 = +10 – 1 Big Friend formula to do your calculations.** ***Now to do +10 you do not have enough beads so, GET HELP from small friend formula and do +10.*** **Step 1: Add 10** = **Step A: Add 50** – Move the heaven bead down to touch the beam on the tens rod. **Step B: Minus 40** – Move all four earth beads down to touch the frame on the tens rod. ***Now we need to finish the formula by doing – 1 on the ones rod, but we do not have enough earth beads to do –1 so, GET HELP from small friend formula and do – 1.*** **Step 2: Minus 1** = **Step A: Minus 5** – Move the heaven bead up to touch the frame on the ones rod. **Step B: Add 4** – Move all four earth beads up to touch the beam on the ones rod. ***(When we do + 10 and – 1, we get to keep 9 in our game.)*** **+10 – 1 = +9**

EXAMPLE: 4

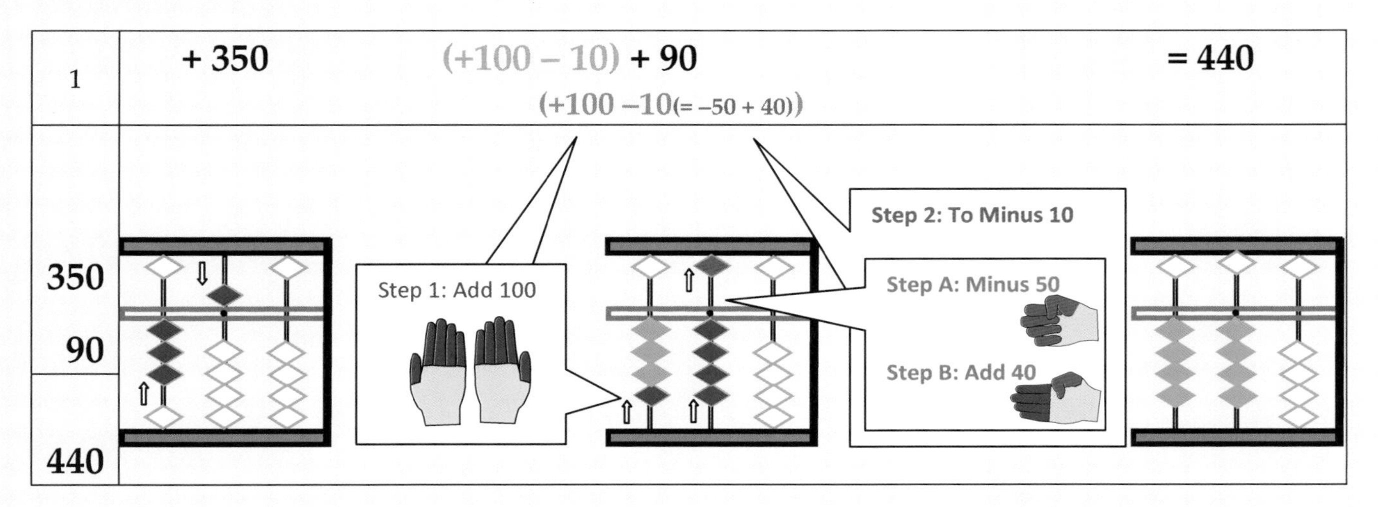

Problem	Action
+ 350	Move **three earth beads** to touch the beam on the **hundreds rod.** Move **heaven bead** to touch the beam on the **tens rod.** There is nothing to do on the ones rod because ones place number is zero.
+ 90	Now we need to +90 on the tens rod, however we do not have enough earth beads to +90 on the tens rod. **When you want to +90 and you do not have enough beads:** **Use +90 = +100 – 10 Big Friend formula to do your calculations.** *(This can be taught as using the same bead movement as for +9, but add on the hundreds rod and minus on the tens rod.)* **Step 1: Add 100** – Move **one earth bead up** to touch the beam on the **hundreds rod.** ***Now we need to finish the formula by doing – 10 on the tens rod, but we do not have enough earth bead to do –10 so, GET HELP from small friends formula and do – 10.*** **Step 2: Minus 10** = **Step A: Minus 50** – Move the heaven bead up to touch the frame on the tens rod. **Step B: Add 40** – Move all four earth beads up to touch the beam on the tens rod. There is nothing to do on the ones rod because the ones place number is zero

ATTENTION

- Ask students to say the formula while they use it. This makes it easy for them to understand and follow through with all the steps in the formula.
- **Do not** say the small friend formula even when you are getting help from it. For instance, in example 4, JUST SAY "– 10" while performing the small friend formula bead movements. DO NOT say the Small friend formula out loud.

EXAMPLE: 5 **+ 90 = + 100 – 10**

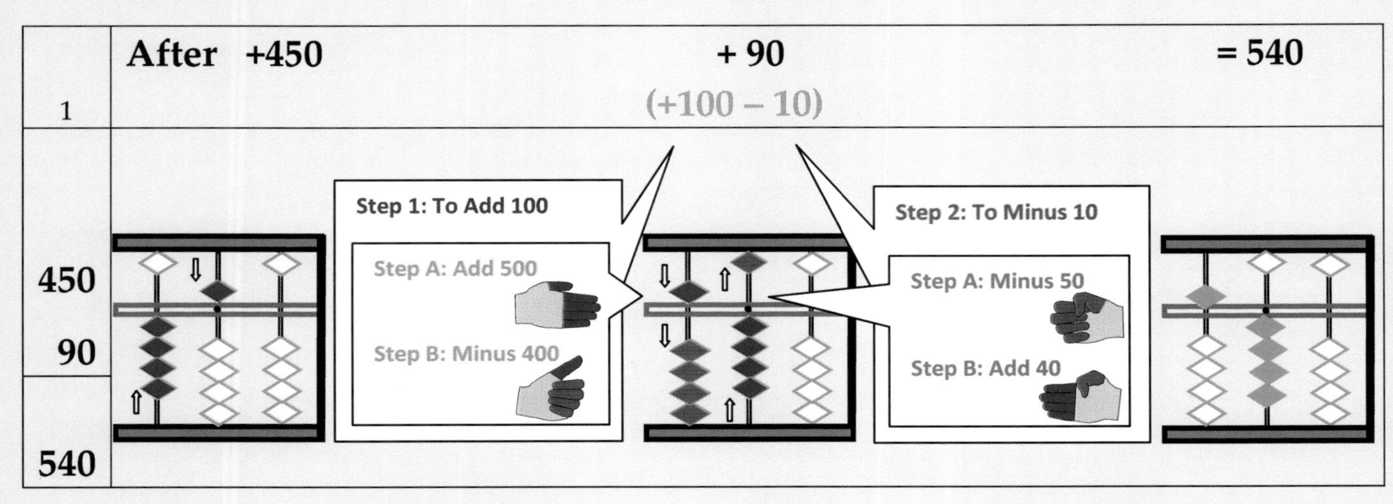

Problem	Action
+ 450	Move **four earth beads** to touch the beam on the **hundreds rod.** Move the **heaven bead down** to touch the beam on the **tens rod**. There is nothing to do on the ones rod because ones place number is zero.
+ 90	Now we need to +90 on the tens rod, however we do not have enough beads on the tens rod to +90. So, now we need to make use of the fact that **90 + 10 = 100** **When you want to +90 and you do not have enough beads:** **Use +90 = +100 – 10 Big Friend formula to do your calculations.** ***Now to do +100 you do not have enough beads so, GET HELP from small friend formula and do +100.*** **Step 1: Add 100** = Step A: Add 500 – Move the heaven bead down to touch the beam on the hundreds rod. Step B: Minus 400 – Move all four earth beads down to touch the frame on the hundreds rod. ***Now we need to finish the formula by doing – 10 on the tens rod, but we do not have enough earth beads to do –10 so, GET HELP from small friend formula and do – 10.*** **Step 2: Minus 10** = Step A: Minus 50 – Move the heaven bead up to touch the frame on the tens rod. Step B: Add 40 – Move all four earth beads up to touch the beam on the tens rod. ***(When we do + 100 and – 10, we get to keep 90 in our game.)*** There is nothing to do on the ones rod because the ones place number is zero.

LESSON 3 – SAMPLE PROBLEMS

+ 9 = + 10 – 1 **+ 90 = + 100 – 10**

TO INTRODUCE –1 OR –10 BY GETTING HELP FROM SMALL FRIENDS
FORMULA OF –1 = –5 +4 OR –10 = –50 +40

Work these problems a few times to study and understand the concept and the relationship between the beads moved.

1	2	3	4	5	6	7	8	9	10	3:S:1 © SAI Speed Math Academy, USA
06	04	25	55	05	14	42	46	44	41	
- 01	01	- 01	- 11	09	11	13	19	11	13	
- 01	- 01	09	90	09	09	09	09	90	91	

TO INTRODUCE +10 OR +100 BY GETTING HELP FROM SMALL FRIENDS
FORMULA OF +10 = +50 –40 OR +100 = +500 –400

Work these problems a few times to study and understand the concept and the relationship between the beads moved.

1	2	3	4	5	6	7	8	9	10	3:S:2 © SAI Speed Math Academy, USA
	45	43	56	46	45	35	56	350	355	
44	10	01	90	09	99	19	99	90	190	
11	09	09	09	09	09	90	99	90	99	

1	2	3	4	5	6	7	8	9	10	3:S:3 © SAI Speed Math Academy, USA
56	24	35	63	59	23	45	155	265	59	
19	29	19	91	99	42	09	199	109	199	
19	29	29	91	- 34	99	91	199	- 144	99	

POINTS TO REMEMBER

The rows above consist of sample problems to introduce this week's formula. Explain to your child when and how to use the formula. Work with the sample problems until your child understands the formula and that the formulas are to be used ONLY when there are not enough beads to add or subtract.

WEEK 4 – LESSON 4 – INTRODUCING + 8 CONCEPT

LESSON 4 – EXAMPLE

CONCEPTS OF THE WEEK

TO ADD = ADD 10, LESS BIG FRIEND **+ 8 = + 10 – 2** **+ 80 = + 100 – 20**

EXAMPLE: 1

1	After	+ 04	+ 08 (+10 – 2)	= 12
04 08 12	ABACUS LOOKS LIKE		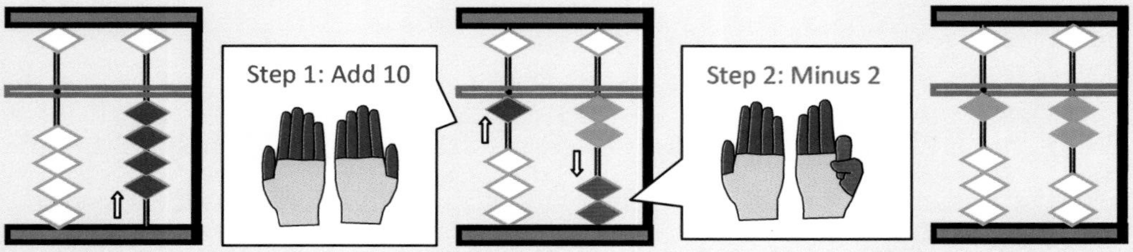 Step 1: Add 10 / Step 2: Minus 2	

Problem	Action
+ 04	There is nothing to do on the tens rod because tens place number is zero. Move all **four earth beads** to touch the beam on the **ones rod.**
+ 08	There is nothing to do on the tens rod because tens place number is zero. Now we need to +8 on the ones rod, however we do not have enough beads on the ones rod to +8. (*The heaven bead is not in the game, but it does not have eight in it, so we cannot get help from it.)* So, now we need to make use of the fact that **8 + 2 = 10** **When you want to +8 and you do not have enough beads:** **Use +8 = +10 – 2 Big Friend formula to do your calculations.** **Step 1: Add 10** – Move **one earth bead up** to touch the beam on the **tens rod**. *(One earth bead on the tens rod is equal to '10')* *We know that there is an eight in the ten (8 + 2 = 10), so let us get help from 10 by adding it to our game. However, we were supposed to +8, instead we did +10, which means we have 2 more than what we need. So, now we have to send the 2 away from our game.* **Step 2: Minus 2** – Move **two earth beads away** from the beam on the **ones rod**. ***(When we do + 10 and – 2, we get to keep 8 in our game.)*** **+10 – 2 = +8**

EXAMPLE: 2 **+ 80 = + 100 – 20**

1	After	+ 70	(+100 – 20) + 80	= 150
70 80 150	ABACUS LOOKS LIKE 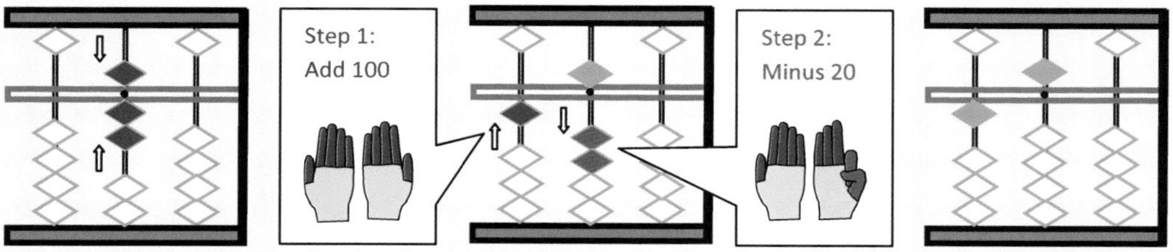		Step 1: Add 100 / Step 2: Minus 20	

Problem	Action
+ 70	Move **two earth beads and the heaven bead** to touch the beam on the **tens rod.** There is nothing to do on the ones rod because ones place number is zero.
+ 80	Now we need to +80 on the tens rod, however we do not have enough earth beads to +80 on the tens rod. **When you want to +80 and you do not have enough beads:** **Use +80 = +100 – 20 Big Friend formula to do your calculations.** *(This can be taught as using the same bead movement as for +8, but add on the hundreds rod and minus on the tens rod.)* **Step 1: Add 100** – Move **one earth bead up** to touch the beam on the **hundreds rod**. **Step 2: Minus 20** – Move **two earth beads away** from the beam on the **tens rod**. There is nothing to do on the ones rod because ones place number is zero.

ATTENTION

- Ask students to say the formula while they use it. This makes it easy for them to understand and follow through with all the steps in the formula.
- When adding 18 to another number (where they need to use +8 formula) students will +10 once and then –2. Make sure they understand that they have to +10 once for the ten in the 18 and **another** +10 to use the formula before finishing with –2 while +8. Ex: 54 + 18
- When following the formula on the tens rod, students usually +100 but then get confused and try to do –2 instead of –20. Make them understand that big friend of 80 is 20 and they help each other.

SAMPLE PROBLEMS

TO INTRODUCE +8 = + 10 – 2 FORMULA

Work these problems a few times to study and understand the concept and the relationship between the beads moved.

1	2	3	4	5	6	7	8	9	10
		04	09	12		22	08	49	44
02	03	08	18	08	17	18	18	18	18
08	08	08	08	28	18	18	09	18	18

4:S:1

© SAI Speed Math Academy, USA

TO INTRODUCE +80 = + 100 – 20 FORMULA

Work these problems a few times to study and understand the concept and the relationship between the beads moved.

1	2	3	4	5	6	7	8	9	10
40	20	30	40	77	122	244	143	99	140
80	80	80	80	80	80	80	80	88	280
80	08	80	180	08	88	88	88	88	180

4:S:2

© SAI Speed Math Academy, USA

1	2	3	4	5	6	7	8	9	10
35	75	27	15	62	95	79	44	26	55
58	83	58	19	- 21	- 62	18	98	81	99
- 21	90	89	08	80	88	- 22	89	18	18
88	18	- 44	17	08	99	89	25	- 12	81

4:S:3

© SAI Speed Math Academy, USA

POINTS TO REMEMBER

The rows above consist of sample problems to introduce this week's formula. Explain to your child when and how to use the formula. Work with the sample problems until your child understands the formula and that the formulas are to be used ONLY when there are not enough beads to add or subtract.

WEEK 5 – LESSON 5 – COMPLETING + 8 USING SMALL FRIENDS FORMULA

CONCEPTS OF THE WEEK

TO ADD = ADD 10, LESS BIG FRIEND **+ 8 = + 10 – 2** **+ 80 = + 100 – 20**

+8 = +10 –2	**+80 = +100 –20**
GET HELP FROM	**GET HELP FROM**
+10 = +50 –40	**+100 = +500 –400**
–2 = –5 +3	**–20 = –50 +30**

In the previous lesson, all the problems where you had to use the formula were simple and you were able to follow the +8 formula directly. This week, we will see how we might have to get help from our LEVEL – 1 Small Friends formula of +10 and – 2 to do +8. Students will understand this week's concept if you make them understand that they are "getting help" from LEVEL 1 formulas to do +8. You may choose to give a quiz/dictation on lessons nine and twelve in LEVEL 1 before introducing them to this week's lesson.

Example 1: 44 + 08 = 52
Set 44 on the abacus. Now to +8 we need to follow the formula of **+8 = +10 – 2**, but we cannot do +10 directly. So to do +10, you have to follow **+10 = +50 – 40** formula from LEVEL – 1.

Example 2: 35 + 08 = 43
Set 35 on the abacus. Now to +8 we need to follow the formula of **+8 = +10** –2. Here, you can directly +10 but you do not have enough ones bead on the ones rod to do –2. So to do –2, you get help from the –2 = –5 + 3 formula from LEVEL – 1.

Example 3: 46 + 08 = 54
Set 46 on the abacus. Here too we need to use the **+8 =** +10 – 2 formula. However, both +10 and – 2 cannot be done directly. To do both these steps you have to get help from LEVEL – 1 Small Friends formulas for +10 = +50 – 40 **and** – 2 = –5 + 3.

Example 4: 350 + 80 = 430
Set 350 on the abacus. To do +80 we now need to use the formula **+80 = +100** –20. Here, to complete the steps you will have to get help from LEVEL – 1 Small Friends formulas for –20 = –50 + 30.

LESSON 5 – EXAMPLE

EXAMPLE: 1

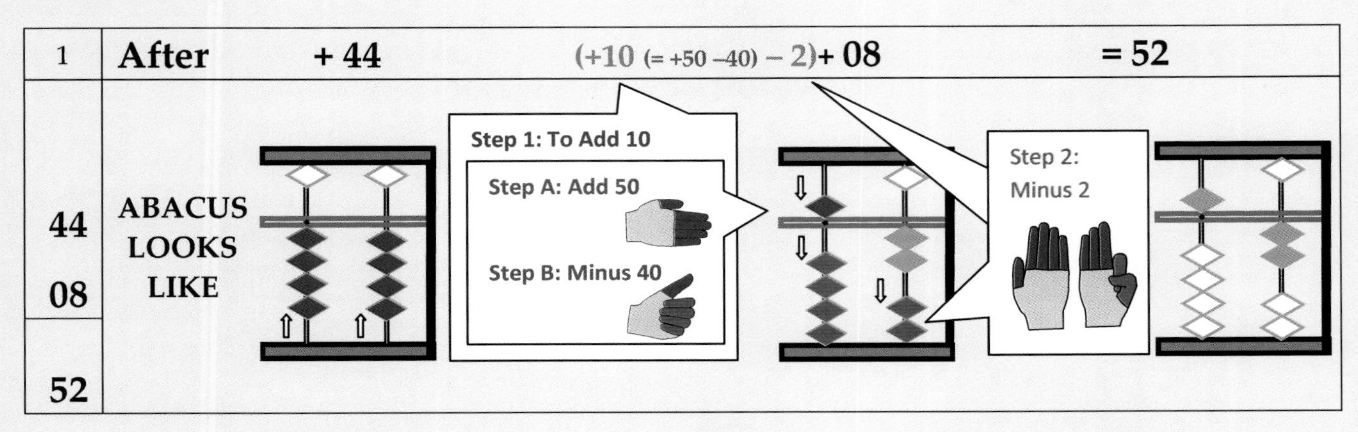

Problem	Action
+ 44	Move all **four earth beads** to touch the beam on the **tens rod.** Move all **four earth beads** to touch the beam on the **ones rod.**
+ 08	There is nothing to do on the tens rod because the tens place number is zero. Now we need to +8 on the ones rod, however we do not have enough beads on the ones rod to +8. (*The heaven bead is not in the game, but it does not have eight in it, so we cannot get help from it.*) So, now we need to make use of the fact that **8 + 2 = 10** **When you want to +8 and you do not have enough beads:** **Use +8 = +10 – 2 Big Friend formula to do your calculations.** *Now to do +10 you do not have enough earth beads so, GET HELP from small friend formula and do +10.* **Step 1: Add 10** = **Step A: Add 50** – Move the heaven bead down to touch the beam on the tens rod. **Step B: Minus 40** – Move all four earth beads away from the beam on the tens rod. **Now complete the +8 formula by doing – 2 on the ones rod.** **Step 2: Minus 2** – Move **two earth beads down** to touch the frame on the **ones rod.** *(When we do + 10 and – 2, we get to keep 8 in our game.)* **+10 – 2 = +8**

EXAMPLE: 2

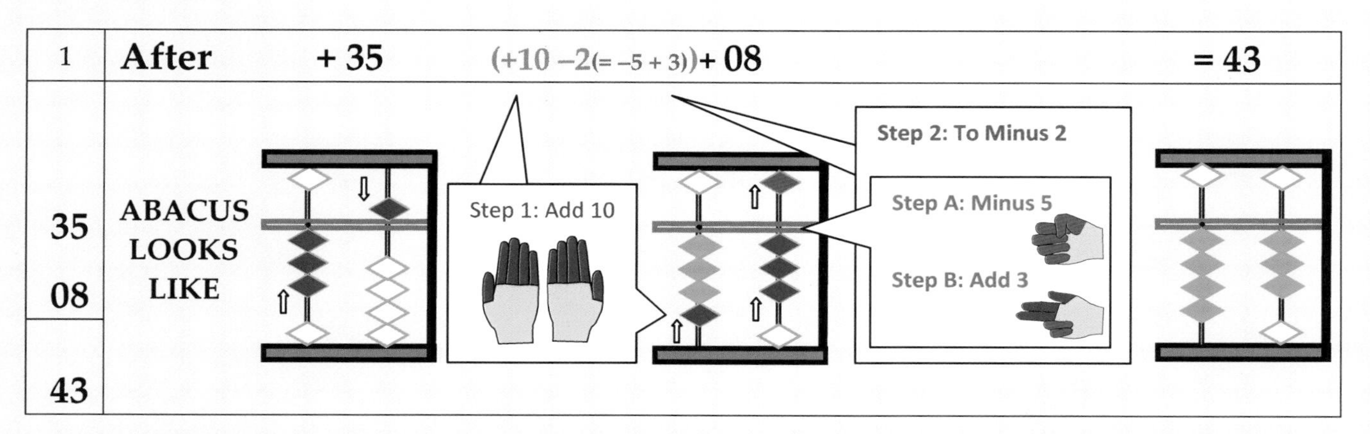

Problem	Action
+ 35	Move **three earth beads** to touch the beam on the **tens rod.** Move the **heaven bead** to touch the beam on the **ones rod.**
+ 08	There is nothing to do on the tens rod because the tens place number is zero. Now we need to +8 on the ones rod, however we do not have enough beads on the ones rod to +8. So, now we need to make use of the fact that **8 + 2 = 10** **When you want to +8 and you do not have enough beads:** **Use +8 = +10 – 2 Big Friend formula to do your calculations.** **<u>Step 1: Add 10</u>** – Move **one earth bead up** to touch the beam on the **tens rod.** *(One earth bead on the tens rod is equal to '10')* ***Now we need to finish the formula by doing – 2 on the ones rod, but we do not have earth beads to do –2 so, GET HELP from small friend formula and do – 2.*** **<u>Step 2: Minus 2</u> =** **<u>Step A: Minus 5</u>** – Move the heaven bead up to touch the frame on the ones rod. **<u>Step B: Add 3</u>** – Move three earth beads up to touch the beam on the ones rod. ***(When we do + 10 and – 2, we get to keep 8 in our game.)*** **+10 – 2 = +8**

EXAMPLE: 3

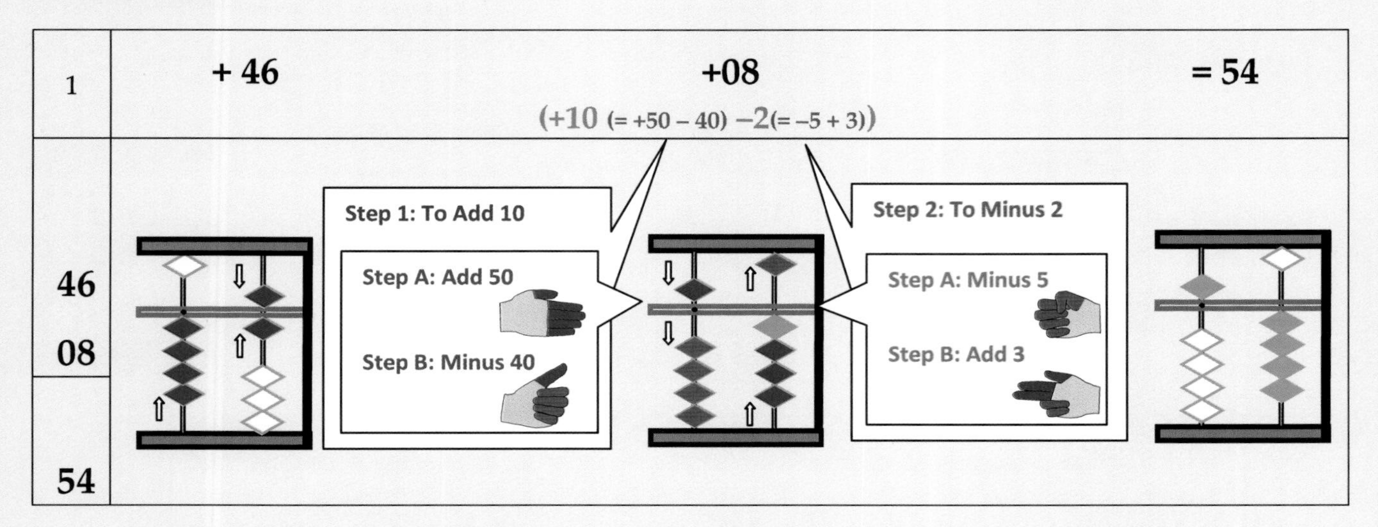

Problem	Action
+ 46	Move **four earth beads** to touch the beam on the **tens rod.** Move the **heaven bead and one earth bead** to touch the beam on the **ones rod.**
+ 08	There is nothing to do on the tens rod because the tens place number is zero. Now we need to +8 on the ones rod, however we do not have enough beads on the ones rod to +8. So, now we need to make use of the fact that **8 + 2 = 10** **When you want to +8 and you do not have enough beads:** **Use +8 = +10 – 2 Big Friend formula to do your calculations.** *Now to do +10 you do not have enough beads so, GET HELP from small friend formula and do +10.* **Step 1: Add 10** = **Step A: Add 50** – Move the heaven bead down to touch the beam on the tens rod. **Step B: Minus 40** – Move all four earth beads down to touch the frame on the tens rod. *Now we need to finish the formula by doing – 2 on the ones rod, but we do not have enough earth beads to do –2 so, GET HELP from small friend formula and do – 2.* **Step 2: Minus 2** = **Step A: Minus 5** – Move the heaven bead up to touch the frame on the ones rod. **Step B: Add 3** – Move three earth beads up to touch the beam on the ones rod. *(When we do + 10 and – 2, we get to keep 8 in our game.)* **+10 – 2 = +8**

EXAMPLE: 4

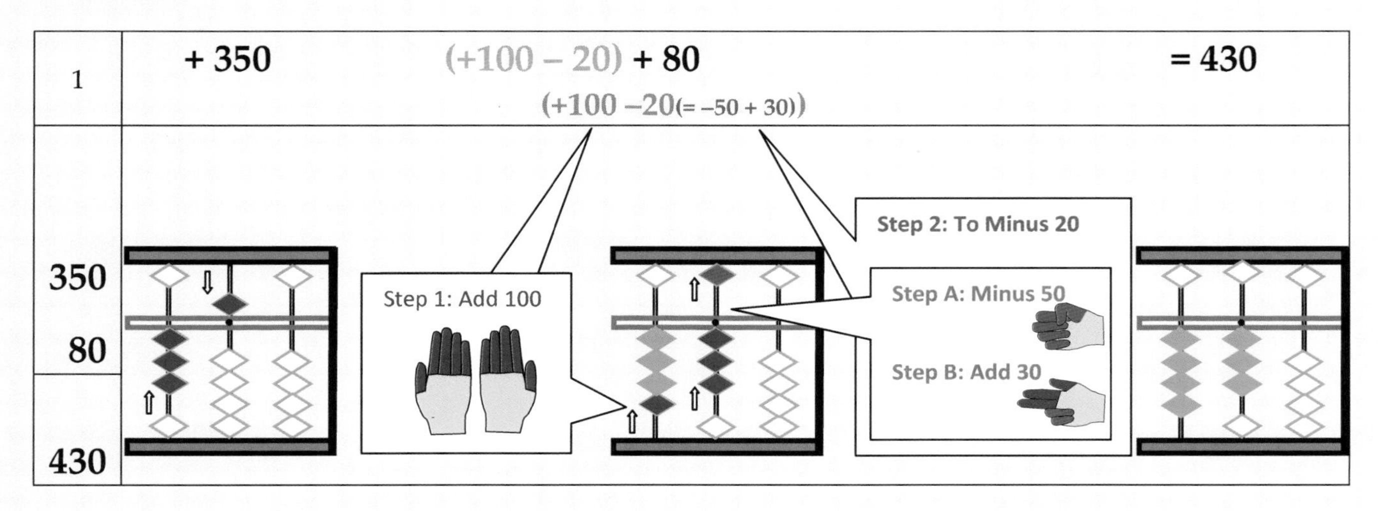

Problem	Action
+ 350	Move **three earth beads** to touch the beam on the **hundreds rod.** Move the **heaven bead** to touch the beam on the **tens rod.** There is nothing to do on the ones rod because ones place number is zero.
+ 80	Now we need to +80 on the tens rod, however we do not have enough earth beads to +80 on the tens rod. **When you want to +80 and you do not have enough beads:** **Use +80 = +100 – 20 Big Friend formula to do your calculations.** *(This can be taught as using the same bead movement as for +8, but add on the hundreds rod and minus on the tens rod.)* **Step 1: Add 100** – Move **one earth bead up** to touch the beam on the **hundreds rod.** ***Now we need to finish the formula by doing – 20 on the tens rod, but we do not have enough earth beads to do –20 so, GET HELP from small friends formula and do – 20.*** **Step 2: Minus 20** = **Step A: Minus 50** – Move the heaven bead up to touch the frame on the tens rod. **Step B: Add 30** – Move three earth beads up to touch the beam on the tens rod. There is nothing to do on the ones rod because the ones place number is zero

ATTENTION

- Ask students to say the formula while they use it. This makes it easy for them to understand and follow through with all the steps in the formula.
- **Do not** say the small friend formula even when you are getting help from it.
- For instance, in example 4, JUST SAY "– 20" while performing the small friend formula bead movements. DO NOT say the Small friend formula out loud.

EXAMPLE: 5 **+ 80 = + 100 – 20**

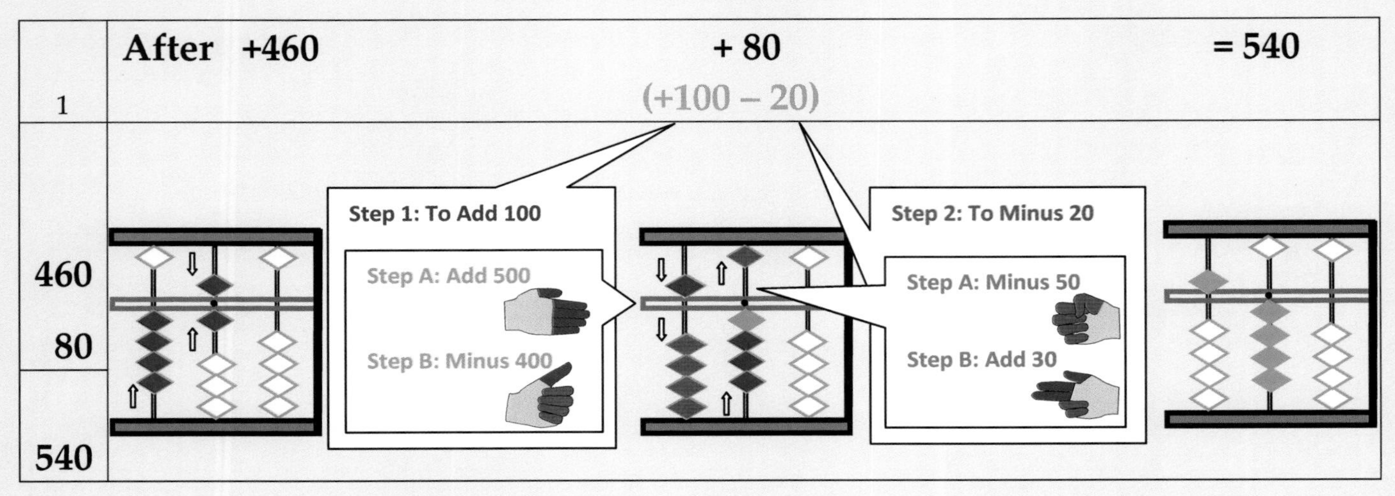

Problem	Action
+ 460	Move **four earth beads** to touch the beam on the **hundreds rod.** Move the **heaven bead** and **one earth** bead touch the beam on the **tens rod**. There is nothing to do on the ones rod because ones place number is zero.
+ 80	Now we need to +80 on the tens rod, however we do not have enough beads on the tens rod to +80. So, now we need to make use of the fact that **80 + 20 = 100** **When you want to +80 and you do not have enough beads:** **Use +80 = +100 – 20 Big Friend formula to do your calculations.** ***Now to do +100 you do not have enough beads so, GET HELP from small friend formula and do +100.*** **Step 1: Add 100** = Step A: Add 500 – Move the heaven bead down to touch the beam on the hundreds rod. Step B: Minus 400 – Move all four earth beads down to touch the frame on the hundreds rod. *Now we need to finish the formula by doing – 20 on the tens rod, but we do not have enough earth beads to do –20 so, GET HELP from small friend formula and do – 20.* **Step 2: Minus 20** = **Step A: Minus 50** – Move the heaven bead up to touch the frame on the tens rod. **Step B: Add 30** – Move three earth beads up to touch the beam on the tens rod. ***(When we do + 100 and – 20, we get to keep 80 in our game.)*** There is nothing to do on the ones rod because the ones place number is zero.

LESSON 5 – SAMPLE PROBLEMS

+ 8 = + 10 – 2 **+ 80 = + 100 – 20**

TO INTRODUCE –2 OR –20 BY GETTING HELP FROM SMALL FRIENDS

FORMULA OF –2 = –5 +3 OR –20 = –50 +30

Work these problems a few times to study and understand the concept and the relationship between the beads moved.

1	2	3	4	5	6	7	8	9	10
07	06	55	66	15	51	75	80	50	45
- 02	- 02	08	08	08	82	18	80	80	18
- 02	08	- 20	08	18	08	80	80	80	80

5:S:1

© SAI Speed Math Academy, USA

TO INTRODUCE +10 OR +100 BY GETTING HELP FROM SMALL FRIENDS

FORMULA OF +10 = +50 –40 OR +100 = +500 –400

Work these problems a few times to study and understand the concept and the relationship between the beads moved.

1	2	3	4	5	6	7	8	9	10
16	25	43	47	45	18	64	56	55	255
18	18	08	18	88	18	82	90	98	280
80	08	80	80	18	18	18	18	80	18

5:S:2

© SAI Speed Math Academy, USA

1	2	3	4	5	6	7	8	9	10
68	56	19	15	45	38	155	241	966	244
19	- 22	38	18	29	21	98	224	- 741	318
88	18	88	19	81	88	98	80	28	- 120
18	89	- 42	83	- 23	- 14	98	08	280	80

5:S:3

© SAI Speed Math Academy, USA

POINTS TO REMEMBER

The rows above consist of sample problems to introduce this week's formula. Explain to your child when and how to use the formula. Work with the sample problems until your child understands the formula and that the formulas are to be used ONLY when there are not enough beads to add or subtract.

WEEK 6 – LESSON 6 – INTRODUCING + 7 CONCEPT

LESSON 6 – EXAMPLE

CONCEPTS OF THE WEEK

TO ADD = ADD 10, LESS BIG FRIEND **+ 7 = + 10 – 3** **+ 70 = + 100 – 30**

EXAMPLE: 1

1	**After** **+ 24**	**+ 07** (+10 – 3)	**= 31**
24 **07** **31**	**ABACUS LOOKS LIKE** 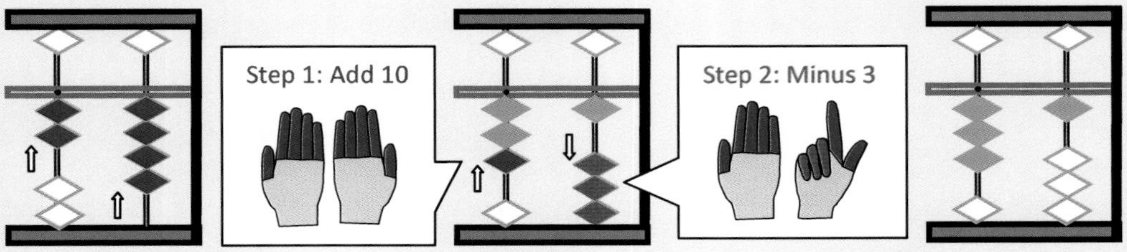	Step 1: Add 10	Step 2: Minus 3

Problem	Action
+ 24	Move **two earth beads** to touch the beam on the **tens rod.** Move all **four earth beads** to touch the beam on the **ones rod.**
+ 07	There is nothing to do on the tens rod because tens place number is zero. Now we need to +7 on the ones rod, however we do not have enough beads on the ones rod to +7. (*The heaven bead is not in the game, but it does not have seven in it, so we cannot get help from it.*) So, now we need to make use of the fact that **7 + 3 = 10** **When you want to +7 and you do not have enough beads:** **Use +7 = +10 – 3 Big Friend formula to do your calculations.** **Step 1: Add 10** – Move **one earth bead up** to touch the beam on the **tens rod.** *(One earth bead on the tens rod is equal to '10')* *We know that there is a seven in the ten (7 + 3 = 10), so let us get help from 10 by adding it to our game. However, we were supposed to +7, instead we did +10, which means we have 3 more than what we need. So, now we have to send the 3 away from our game.* **Step 2: Minus 3** – Move **three earth beads away** from the beam on the **ones rod.** ***(When we do + 10 and – 3, we get to keep 7 in our game.)*** **+10 – 3 = +7**

EXAMPLE: 2 **+ 70 = + 100 – 30**

1	**After** **+ 90** (+100 – 30) **+ 70** **= 160**
90 **70** **160**	**ABACUS LOOKS LIKE** 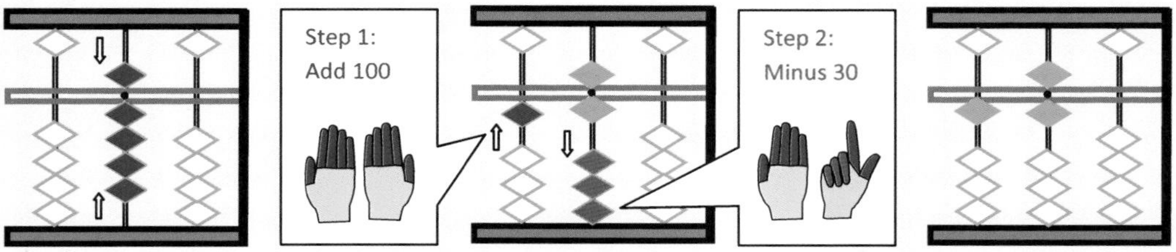

Problem	Action
+ 90	Move **four earth beads and heaven bead** to touch the beam on the **tens rod.** There is nothing to do on the ones rod because ones place number is zero.
+ 70	Now we need to +70 on the tens rod, however we do not have enough earth beads to +70 on the tens rod. **When you want to +70 and you do not have enough beads:** **Use +70 = +100 – 30 Big Friend formula to do your calculations.** *(This can be taught as using the same bead movement as for +7, but add on the hundreds rod and minus on the tens rod.)* **Step 1: Add 100** – Move **one earth bead up** to touch the beam on the **hundreds rod**. **Step 2: Minus 30** – Move **three earth beads down** to touch the frame on the **tens rod**. There is nothing to do on the ones rod because ones place number is zero.

ATTENTION

- Ask students to say the formula while they use it. This makes it easy for them to understand and follow through with all the steps in the formula.
- When adding 17 to another number (where they need to use +7 formula) students will +10 once and then **–3**. Make sure they understand that they have to +10 once for the ten in the 17 and **another** +10 to use the formula before finishing with **–3** while +7. Ex: 54 + 17
- When following the formula on the tens rod, students usually +100 but then get confused and try to do –3 instead of –30. Make them understand that big friend of 70 is 30 and they help each other.

LESSON 6 – SAMPLE PROBLEMS

TO INTRODUCE +7 = + 10 – 3 FORMULA

Work these problems a few times to study and understand the concept and the relationship between the beads moved.

1	2	3	4	5	6	7	8	9	10	6:S:1 © SAI Speed Math Academy, USA
				77	88	49	134	144	123	
04	13	18	29	- 33	07	17	07	17	17	
07	17	17	17	17	- 33	- 33	30	17	80	

TO INTRODUCE +70 = + 100 – 30 FORMULA

Work these problems a few times to study and understand the concept and the relationship between the beads moved.

1	2	3	4	5	6	7	8	9	10	6:S:2 © SAI Speed Math Academy, USA
65	53	95	19	44	56	19	80	55	28	
80	87	70	17	70	- 12	27	79	33	17	
79	87	09	78	17	77	70	17	77	70	

1	2	3	4	5	6	7	8	9	10	6:S:3 © SAI Speed Math Academy, USA
14	08	25	53	15	26	47	199	249	244	
07	07	18	87	19	28	- 13	177	377	17	
17	80	77	70	07	27	07	- 133	170	87	
70	70	30	70	18	09	70	117	- 423	77	

POINTS TO REMEMBER

The rows above consist of sample problems to introduce this week's formula. Explain to your child when and how to use the formula. Work with the sample problems until your child understands the formula and that the formulas are to be used ONLY when there are not enough beads to add or subtract.

WEEK 7 – LESSON 7 – COMPLETING + 7 USING SMALL FRIENDS FORMULA

CONCEPTS OF THE WEEK

TO ADD = ADD 10, LESS BIG FRIEND **+ 7 = + 10 – 3** **+ 70 = + 100 – 30**

+7 = +10 –3	**+70 = +100 –30**
GET HELP FROM	**GET HELP FROM**
+10 = +50 –40	**+100 = +500 –400**
–3 = –5 +2	**–30 = –50 +20**

In the previous lesson, all the problems where you had to use the formula were simple and you were able to follow the +7 formula directly. This week, we will see how we might have to get help from our LEVEL – 1 Small Friends formula of +10 and – 3 to do +7. Students will understand this week's concept if you make them understand that they are "getting help" from LEVEL – 1 formulas to do +7. You may choose to give a quiz/dictation on lessons nine and fourteen in LEVEL 1 before introducing them to this week's lesson.

Example 1: 44 + 07 = 51
Set 44 on the abacus. Now to +7 we need to follow the formula of **+7 = +10 – 3**, but we cannot do +10 directly. So to do +10, you have to follow **+10 = +50 – 40** formula from LEVEL – 1.

Example 2: 27 + 07 = 44
Set 27 on the abacus. Now to +7 we need to follow the formula of **+7 = +10** –3. Here, you can directly +10 but you do not have enough ones bead on the ones rod to do –3. So to do –3, you get help from the –3 = –5 + 2 formula from LEVEL – 1.

Example 3: 45 + 07 = 52
Set 45 on the abacus. Here too we need to use the **+7 = +10 – 3** formula. However, both +10 and – 3 cannot be done directly. To do both these steps you have to get help from LEVEL – 1 Small Friends formulas for **+10 = +50 – 40 and – 3 = –5 + 2.**

Example 4: 160 + 70 = 230
Set 160 on the abacus. To do +70 we now need to use the formula **+70 = +100** –30. Here, to complete the steps you will have to get help from LEVEL – 1 Small Friends formulas for –30 = –50 + 20.

LESSON 7 – EXAMPLE

EXAMPLE: 1

1	**After** **+ 44** (+10 (= +50 –40) – 3)**+ 07** **= 51**
44 **07** **51**	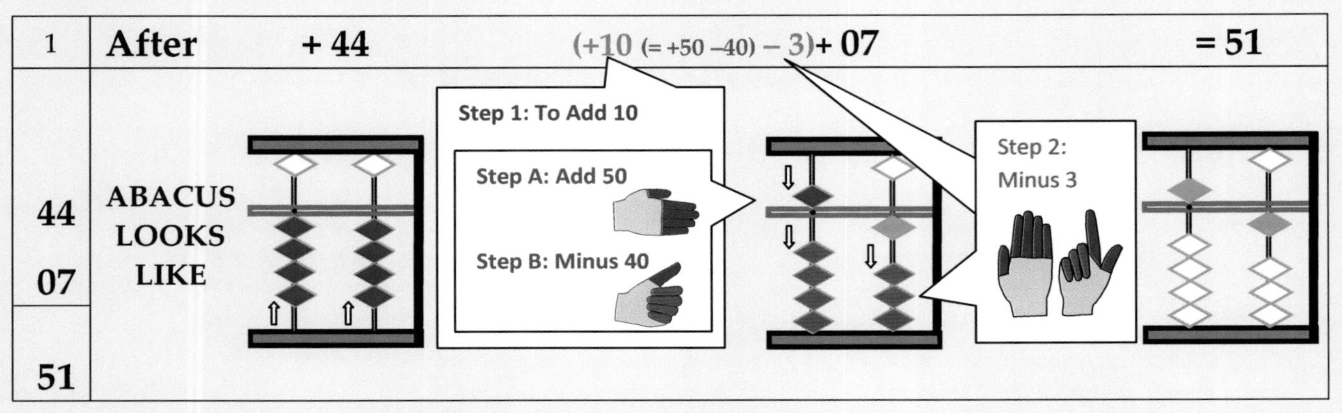

Problem	Action
+ 44	Move all **four earth beads** to touch the beam on the **tens rod.** Move all **four earth beads** to touch the beam on the **ones rod.**
+ 07	There is nothing to do on the tens rod because the tens place number is zero. Now we need to +7 on the ones rod, however we do not have enough beads on the ones rod to +7. (*The heaven bead is not in the game, but it does not have seven in it, so we cannot get help from it.)* So, now we need to make use of the fact that **7 + 3 = 10** **When you want to +7 and you do not have enough beads:** **Use +7 = +10 – 3 Big Friend formula to do your calculations.** ***Now to do +10 you do not have enough earth beads so, GET HELP from small friend formula and do +10.*** **Step 1: Add 10** = **Step A: Add 50** – Move the heaven bead down to touch the beam on the tens rod. **Step B: Minus 40** – Move all four earth beads down to touch the frame on the tens rod. **Now complete the +7 formula by doing – 3 on the ones rod.** **Step 2: Minus 3** – Move **three earth beads away** from the beam on the **ones rod.** *(When we do + 10 and – 3, we get to keep 7 in our game.)* **+10 – 3 = +7**

EXAMPLE: 2

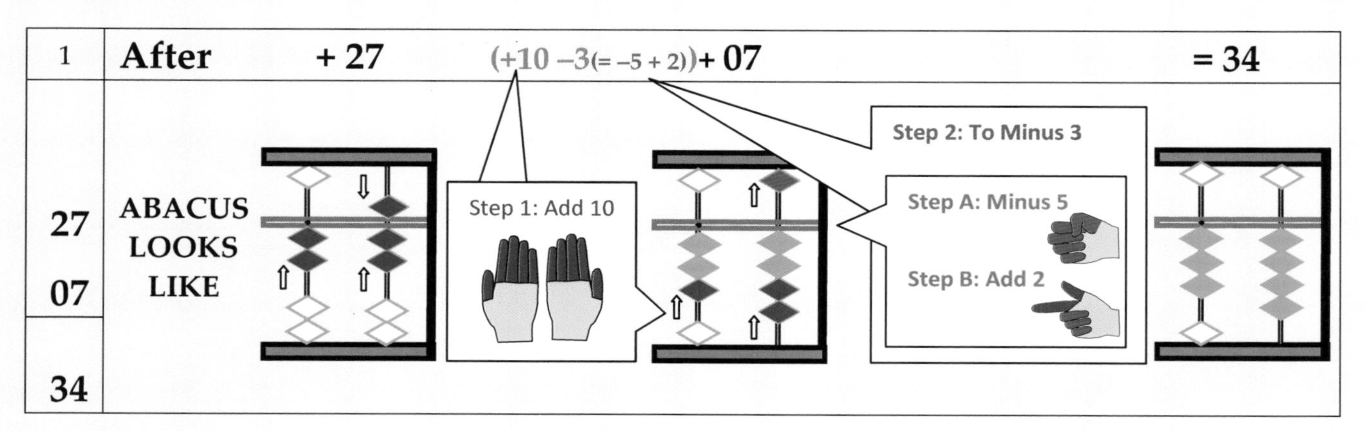

Problem	Action
+ 27	Move **two earth beads** to touch the beam on the **tens rod.** Move the **heaven bead and two earth beads** to touch the beam on the **ones rod.**
+ 07	There is nothing to do on the tens rod because the tens place number is zero. Now we need to +7 on the ones rod, however we do not have enough beads on the ones rod to +7. So, now we need to make use of the fact that **7 + 3 = 10** **When you want to +7 and you do not have enough beads:** **Use +7 = +10 – 3 Big Friend formula to do your calculations.** Step 1: Add 10 – Move **one earth bead up** to touch the beam on the **tens rod.** *(One earth bead on the tens rod is equal to '10')* *Now we need to finish the formula by doing – 3 on the ones rod, however we do not have earth beads to do –3 so, GET HELP from small friend formula and do – 3.* Step 2: Minus 3 = Step A: Minus 5 – Move the heaven bead up to touch the frame on the ones rod. Step B: Add 2 – Move two earth beads up to touch the beam on the ones rod. *(When we do + 10 and – 3, we get to keep 7 in our game.)* **+10 – 3 = +7**

EXAMPLE: 3

1	+ 45	+07 (+10 (= +50 – 40) –3(= –5 + 2))	= 52

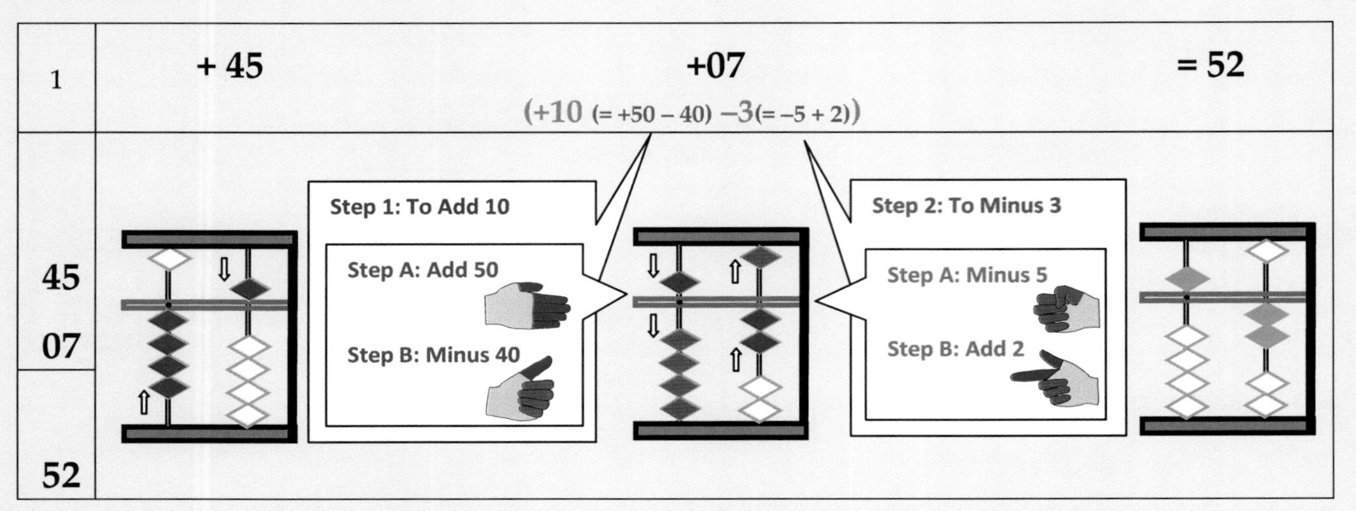

Problem	Action
+ 45	Move **four earth beads** to touch the beam on the **tens rod.** Move the **heaven bead** to touch the beam on the **ones rod.**
+ 07	There is nothing to do on the tens rod because the tens place number is zero. Now we need to +7 on the ones rod, however we do not have enough beads on the ones rod to +7. So, now we need to make use of the fact that 7 **+ 3 = 10** **When you want to +7 and you do not have enough beads:** **Use +7 = +10 – 3 Big Friend formula to do your calculations.** ***Now to do +10 you do not have enough beads so, GET HELP from small friend formula and do +10.*** **Step 1: Add 10** = **Step A: Add 50** – Move the heaven bead down to touch the beam on the tens rod. **Step B: Minus 40** – Move all four earth beads down to touch the frame on the tens rod. ***Now we need to finish the formula by doing – 3 on the ones rod, but we do not have enough earth beads to do –3 so, GET HELP from small friend formula and do – 3.*** **Step 2: Minus 3** = **Step A: Minus 5** – Move the heaven bead up to touch the frame on the ones rod. **Step B: Add 2** – Move two earth beads up to touch the beam on the ones rod. ***(When we do + 10 and – 3, we get to keep 7 in our game.)*** **+10 – 3= +7**

EXAMPLE: 4

1	+ 160	(+100 – 30) + 70 (+100 –30(= –50 + 20))	= 230

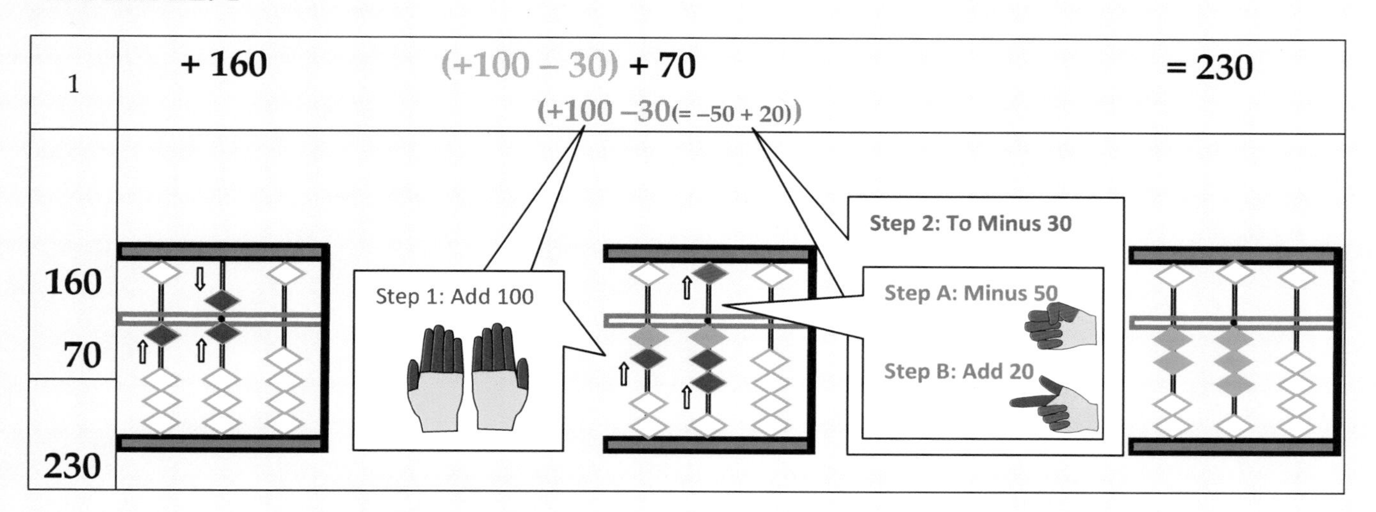

Problem	Action
+ 160	Move **one earth bead** to touch the beam on the **hundreds rod.** Move the **heaven bead and one earth bead** to touch the beam on the **tens rod.** There is nothing to do on the ones rod because ones place number is zero.
+ 70	Now we need to +70 on the tens rod, however we do not have enough earth beads to +70 on the tens rod. **When you want to +70 and you do not have enough beads:** **Use +70 = +100 – 30 Big Friend formula to do your calculations.** *(This can be taught as using the same bead movement as for +7, but add on the hundreds rod and minus on the tens rod.)* Step 1: Add 100 – Move **one earth bead up** to touch the beam on the **hundreds rod.** ***Now we need to finish the formula by doing – 30 on the tens rod, but we do not have enough earth bead to do –30 so, GET HELP from small friends formula and do – 30.*** Step 2: Minus 30 = Step A: Minus 50 – Move the heaven bead up to touch the frame on the tens rod. Step B: Add 20 – Move two earth beads up to touch the beam on the tens rod. There is nothing to do on the ones rod because the ones place number is zero.

ATTENTION

- Ask students to say the formula while they use it. This makes it easy for them to understand and follow through with all the steps in the formula
- **Do not** say the small friend formula even when you are getting help from it.
- For instance, in example 4, JUST SAY "– 30" while performing the small friend formula bead movements. DO NOT say the Small friend formula out loud.

EXAMPLE: 5 **+ 70 = + 100 – 30**

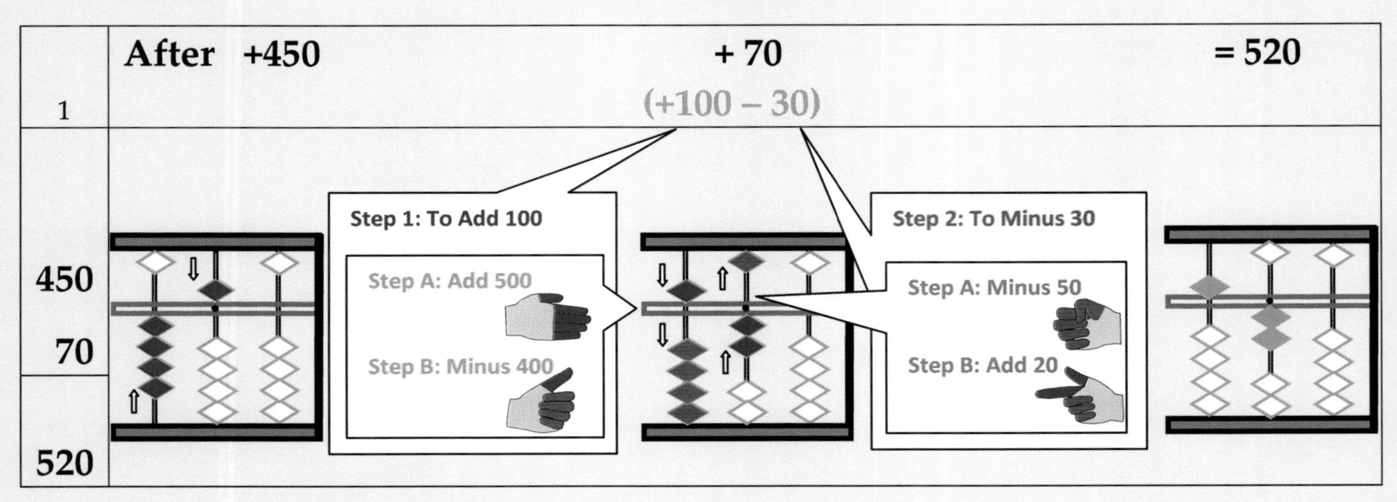

Problem	Action
+ 450	Move **four earth beads** to touch the beam on the **hundreds rod.** Move the **heaven bead** touch the beam on the **tens rod**. There is nothing to do on the ones rod because ones place number is zero.
+ 70	Now we need to +70 on the tens rod, however we do not have enough beads on the tens rod to +70. So, now we need to make use of the fact that **80 + 20 = 100** **When you want to +70 and you do not have enough beads:** **Use +70 = +100 – 30 Big Friend formula to do your calculations.** ***Now to do +100 you do not have enough beads so, GET HELP from small friend formula and do +100.*** **Step 1: Add 100** = **Step A: Add 500** – Move the heaven bead down to touch the beam on the hundreds rod. **Step B: Minus 400** – Move all four earth beads down to touch the frame on the hundreds rod. ***Now we need to finish the formula by doing – 30 on the tens rod, but we do not have enough earth beads to do –30 so, GET HELP from small friend formula and do – 30.*** **Step 2: Minus 30** = **Step A: Minus 50** – Move the heaven bead up to touch the frame on the tens rod. **Step B: Add 20** – Move two earth beads up to touch the beam on the tens rod. ***(When we do + 100 and – 30, we get to keep 70 in our game.)*** There is nothing to do on the ones rod because the ones place number is zero

LESSON 7 – SAMPLE PROBLEMS

+ 7 = + 10 – 3 **+ 70 = + 100 – 30**

TO INTRODUCE –3 OR –30 BY GETTING HELP FROM SMALL FRIENDS
FORMULA OF –3 = –5 +2 OR –30 = –50 +20

Work these problems a few times to study and understand the concept and the relationship between the beads moved.

1	2	3	4	5	6	7	8	9	10	7:S:1
06	25	57	55	48	18	80	52	77	61	
07	07	- 30	17	17	27	70	70	70	71	
07	- 30	07	- 30	17	17	70	07	- 33	77	

© SAI Speed Math Academy, USA

TO INTRODUCE +10 OR +100 BY GETTING HELP FROM SMALL FRIENDS
FORMULA OF +10 = +50 –40 OR +100 = +500 –400

Work these problems a few times to study and understand the concept and the relationship between the beads moved.

1	2	3	4	5	6	7	8	9	10	7:S:2
48	39	36	25	57	45	75	55	55	150	
07	17	17	17	07	07	70	97	78	190	
07	17	03	17	- 33	70	07	70	77	170	

© SAI Speed Math Academy, USA

1	2	3	4	5	6	7	8	9	10	7:S:3
56	87	68	26	46	37	146	186	144	285	
71	78	70	27	77	37	17	270	33	70	
09	97	- 11	90	27	- 20	70	78	- 11	279	
80	- 30	09	- 23	77	41	98	- 333	370	21	

© SAI Speed Math Academy, USA

POINTS TO REMEMBER

The rows above consist of sample problems to introduce this week's formula. Explain to your child when and how to use the formula. Work with the sample problems until your child understands the formula and that the formulas are to be used ONLY when there are not enough beads to add or subtract.

WEEK 9 – LESSON 8 – INTRODUCING + 6 CONCEPT

LESSON 8 – EXAMPLE

CONCEPTS OF THE WEEK
TO ADD = ADD 10, LESS BIG FRIEND   + 6 = + 10 – 4   + 60 = + 100 – 40

EXAMPLE: 1

1	After	+ 04	+ 06 (+10 – 4)	= 10
04 06 10	ABACUS LOOKS LIKE			

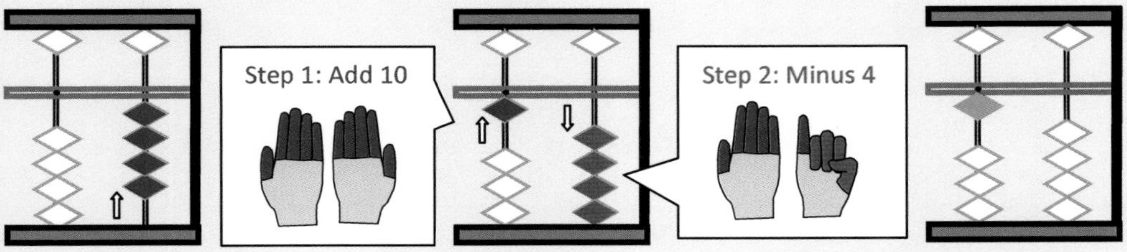

Problem	Action
+ 04	There is nothing to do on the tens rod because tens place number is zero. Move all **four earth beads** to touch the beam on the **ones rod.**
+ 06	There is nothing to do on the tens rod because tens place number is zero. Now we need to +6 on the ones rod, however we do not have enough beads on the ones rod to +6. (*The heaven bead is not in the game, but it does not have six in it, so we cannot get help from it.*) So, now we need to make use of the fact that **6 + 4 = 10** **When you want to +6 and you do not have enough beads:** **Use +6 = +10 – 4 Big Friend formula to do your calculations.** **Step 1: Add 10** – Move **one earth bead up** to touch the beam on the **tens rod**. (*One earth bead on the tens rod is equal to '10')* *We know that there is a six in the ten (6 + 4 = 10), so let us get help from 10 by adding it to our game. However, we were supposed to +6, instead we did +10, which means we have 4 more than what we need. So, now we have to send the 4 away from our game.* **Step 2: Minus 4** – Move **four earth beads away** from the beam on the **ones rod**. ***(When we do + 10 and – 4, we get to keep 6 in our game.)*** **+10 – 4 = +6**

EXAMPLE: 2 **+ 60 = + 100 – 40**

1	After	+ 90	(+100 – 40) + 60	= 150
90 60 150	ABACUS LOOKS LIKE		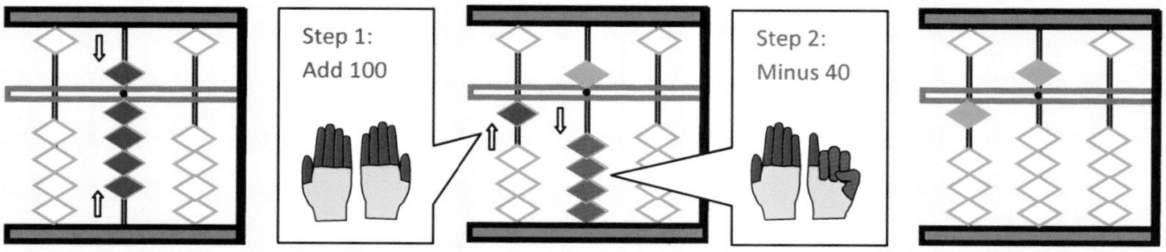 Step 1: Add 100 Step 2: Minus 40	

Problem	Action
+ 90	Move **four earth beads and the heaven bead** to touch the beam on the **tens rod.** There is nothing to do on the ones rod because ones place number is zero.
+ 60	Now we need to +60 on the tens rod, however we do not have enough earth beads to +60 on the tens rod. **When you want to +60 and you do not have enough beads:** **Use +60 = +100 – 40 Big Friend formula to do your calculations.** *(This can be taught as using the same bead movement as for +6, but add on the hundreds rod and minus on the tens rod.)* **Step 1: Add 100** – Move **one earth bead up** to touch the beam on the **hundreds rod**. **Step 2: Minus 40** – Move **four earth beads away** from the beam on the **tens rod**. There is nothing to do on the ones rod because ones place number is zero.

ATTENTION

- Ask students to say the formula while they use it. This makes it easy for them to understand and follow through with all the steps in the formula.
- When adding 16 to another number (where they need to use +6 formula) students will +10 once and then –4. Make sure they understand that they have to +10 once for the ten in the 16 and **another** +10 to use the formula before finishing with –4 while +6. Ex: 54 + 16
- When following the formula on the tens rod, students usually +100 but then get confused and try to do –4 instead of –40. Make them understand that big friend of 60 is 40 and they help each other.

LESSON 8 – SAMPLE PROBLEMS

TO INTRODUCE +6 = + 10 – 4 FORMULA

Work these problems a few times to study and understand the concept and the relationship between the beads moved.

1	2	3	4	5	6	7	8	9	10
		04	09	12			08	49	44
04	09	16	16	07	29	44	41	16	36
06	06	80	09	06	16	16	16	17	- 40

9:S:1

© SAI Speed Math Academy, USA

TO INTRODUCE +60 = + 100 – 40 FORMULA

Work these problems a few times to study and understand the concept and the relationship between the beads moved.

1	2	3	4	5	6	7	8	9	10
40	95	49	95	77	88	99	143	69	194
65	64	60	69	- 33	- 44	66	61	76	66
- 04	06	06	16	66	69	- 44	16	68	- 140

9:S:2

© SAI Speed Math Academy, USA

1	2	3	4	5	6	7	8	9	10
54	23	22	34	33	55	49	43	89	33
06	44	43	26	44	99	16	43	- 66	22
14	27	19	15	- 33	86	34	11	31	- 41
16	66	76	19	66	10	66	60	76	16

9:S:3

© SAI Speed Math Academy, USA

POINTS TO REMEMBER

The rows above consist of sample problems to introduce this week's formula. Explain to your child when and how to use the formula. Work with the sample problems until your child understands the formula and that the formulas are to be used ONLY when there are not enough beads to add or subtract.

WEEK 10 – LESSON 9 – COMPLETING + 6 USING SMALL FRIENDS FORMULA

CONCEPTS OF THE WEEK

TO ADD = ADD 10, LESS BIG FRIEND **+ 6 = + 10 – 4** **+ 60 = + 100 – 40**

+6 = +10 –4	**+60 = +100 –40**
GET HELP FROM	**GET HELP FROM**
+10 = +50 –40	**+100 = +500 –400**
–4 = –5 +1	**–40 = –50 +10**

In the previous lesson, all the problems where you had to use the formula were simple and you were able to follow the +6 formula directly. This week, we will see how we might have to get help from our LEVEL – 1 Small Friends formula of +10 and – 4 to complete +6 formula. Students will understand this week's concept if you make them understand that they are "getting help" from LEVEL 1 formulas to do +6. You may choose to give a quiz/dictation on lessons nine and sixteen in LEVEL 1 before introducing them to this week's lesson.

Example 1: 44 + 06 = 50
Set 44 on the abacus. Now to +6 we need to follow the formula of **+6 = +10 – 4**, but we cannot do +10 directly. So, now to do +10 you have to follow **+10 = +50 – 40** formula from LEVEL – 1.

Example 2: 37 + 06 = 43
Set 37 on the abacus. Now to +6 we need to follow the formula of **+6 = +10** –4. Here, you can directly +10 but you do not have enough ones bead on the ones rod to do –4. So, now to do –4 you get help from the –4 = –5 + 1 formula from LEVEL – 1.

Example 3: 48 + 06 = 54
Set 48 on the abacus. Here too we need to use the **+6 =** +10 – 4 formula. However, both +10 and – 4 cannot be done directly. To do both these steps you have to get help from LEVEL – 1 Small Friends formulas for **+10 = +50 – 40 and** – 4 = –5 + 1**.**

Example 4: 360 + 60 = 420
Set 360 on the abacus. To do +60 we now need to use the formula **+60 = +100** –40. Here, too to complete the steps you will have to get help from LEVEL – 1 Small Friends formulas for –40 = –50 + 10.

LESSON 9 – EXAMPLE

EXAMPLE: 1

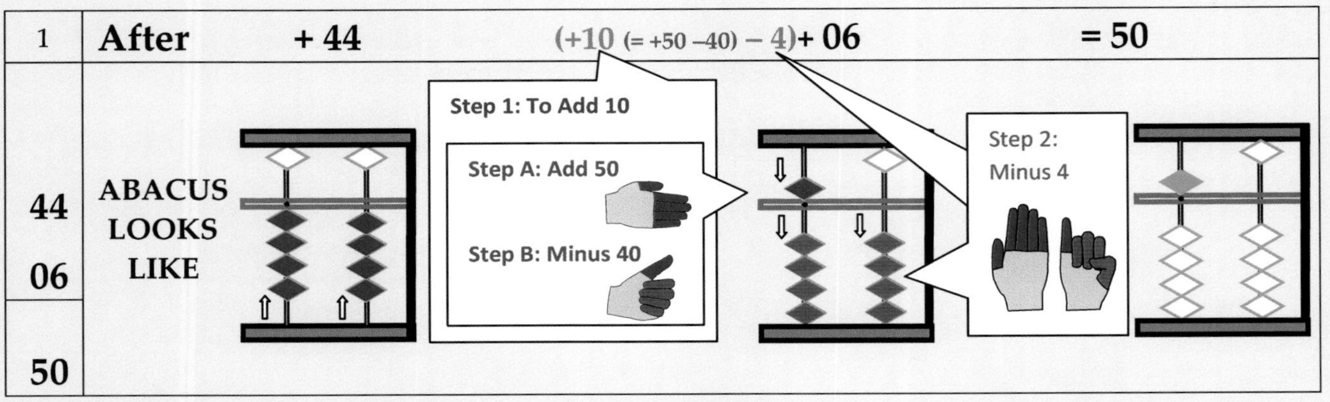

Problem	Action
+ 44	Move all **four earth beads** to touch the beam on the **tens rod.** Move all **four earth beads** to touch the beam on the **ones rod.**
+ 06	There is nothing to do on the tens rod because the tens place number is zero. Now we need to +6 on the ones rod, however we do not have enough beads on the ones rod to +6. (*The heaven bead is not in the game, but it does not have six in it, so we cannot get help from it.*) So, now we need to make use of the fact that **6 + 4 = 10** **When you want to +6 and you do not have enough beads:** **Use +6 = +10 – 4 Big Friend formula to do your calculations.** *Now to do +10 you do not have enough earth beads so, GET HELP from small friend formula and do +10.* **Step 1: Add 10** = **Step A: Add 50** – Move the heaven bead down to touch the beam on the tens rod. **Step B: Minus 40** – Move all four earth beads away from the beam on the tens rod. **Now complete the +6 formula by doing – 4 on the ones rod.** **Step 2: Minus 4** – Move **four earth beads down** to touch the frame on the **ones rod**. *(When we do + 10 and – 4, we get to keep 6 in our game.)* **+10 – 4 = +6**

EXAMPLE: 2

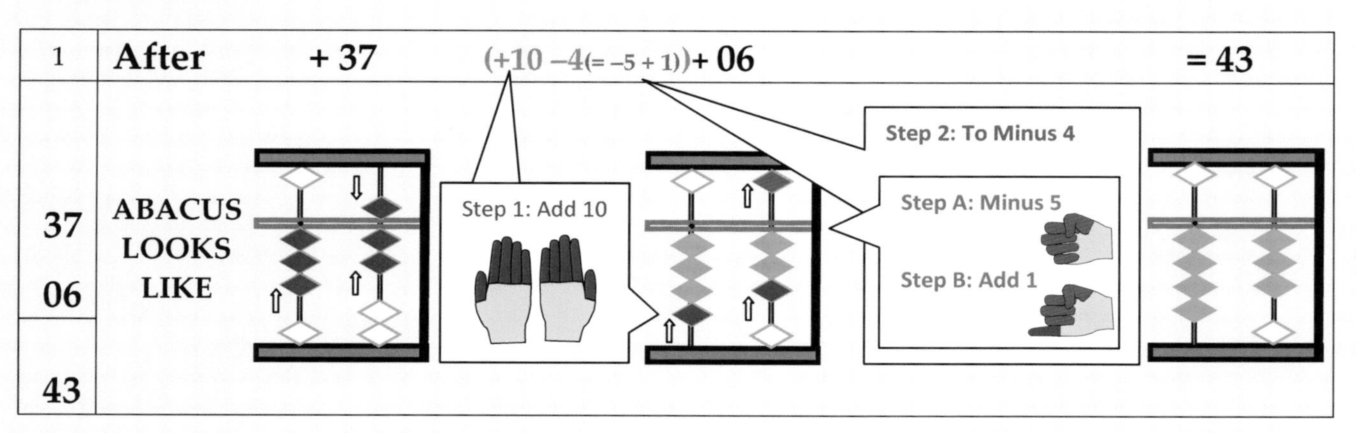

Problem	Action
+ 37	Move **three earth beads** to touch the beam on the **tens rod.** Move the **heaven bead and two earth beads** to touch the beam on the **ones rod.**
+ 06	There is nothing to do on the tens rod because the tens place number is zero. Now we need to +6 on the ones rod, however we do not have enough beads on the ones rod to +6. So, now we need to make use of the fact that **6 + 4 = 10** **When you want to +6 and you do not have enough beads:** **Use +6 = +10 – 4 Big Friend formula to do your calculations.** **Step 1: Add 10** – Move **one earth bead up** to touch the beam on the **tens rod**. *(One earth bead on the tens rod is equal to '10')* ***Now we need to finish the formula by doing – 4 on the ones rod, but we do not have earth beads to do –4 so, GET HELP from small friend formula and do – 4.*** **Step 2: Minus 4** = **Step A: Minus 5** – Move the heaven bead up to touch the frame on the ones rod. **Step B: Add 1** – Move one earth bead up to touch the beam on the ones rod. ***(When we do + 10 and – 4, we get to keep 6 in our game.)*** **+10 – 4 = +6**

EXAMPLE: 3

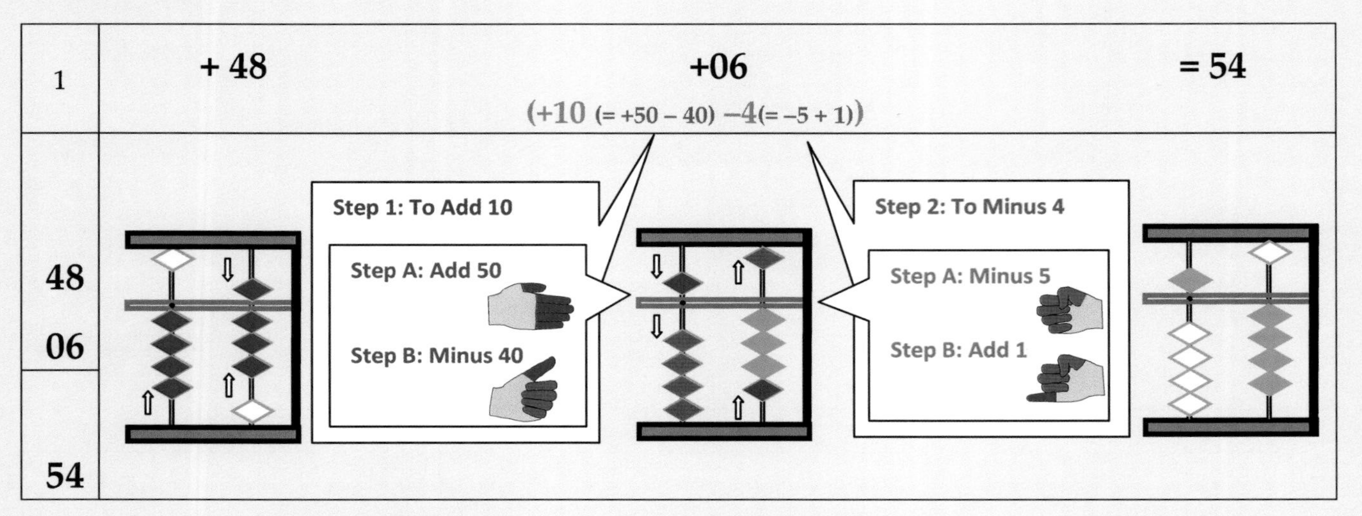

Problem	Action
+ 48	Move **four earth beads** to touch the beam on the **tens rod.** Move the **heaven bead and three earth beads** to touch the beam on the **ones rod.**
+ 06	There is nothing to do on the tens rod because the tens place number is zero. Now we need to +6 on the ones rod, however we do not have enough beads on the ones rod to +6. So, now we need to make use of the fact that **6 + 4 = 10** **When you want to +6 and you do not have enough beads:** **Use +6 = +10 – 4 Big Friend formula to do your calculations.** *Now to do +10 you do not have enough beads so, GET HELP from small friend formula and do +10.* **Step 1: Add 10** = **Step A: Add 50** – Move the heaven bead down to touch the beam on the tens rod. **Step B: Minus 40** – Move all four earth beads down to touch the frame on the tens rod. *Now we need to finish the formula by doing – 4 on the ones rod, but we do not have enough earth beads to do –4 so, GET HELP from small friend formula and do – 4.* **Step 2: Minus 4** = **Step A: Minus 5** – Move the heaven bead up to touch the frame on the ones rod. **Step B: Add 1** – Move one earth bead up to touch the beam on the ones rod. *(When we do + 10 and – 4, we get to keep 6 in our game.)* **+10 – 4 = +6**

EXAMPLE: 4

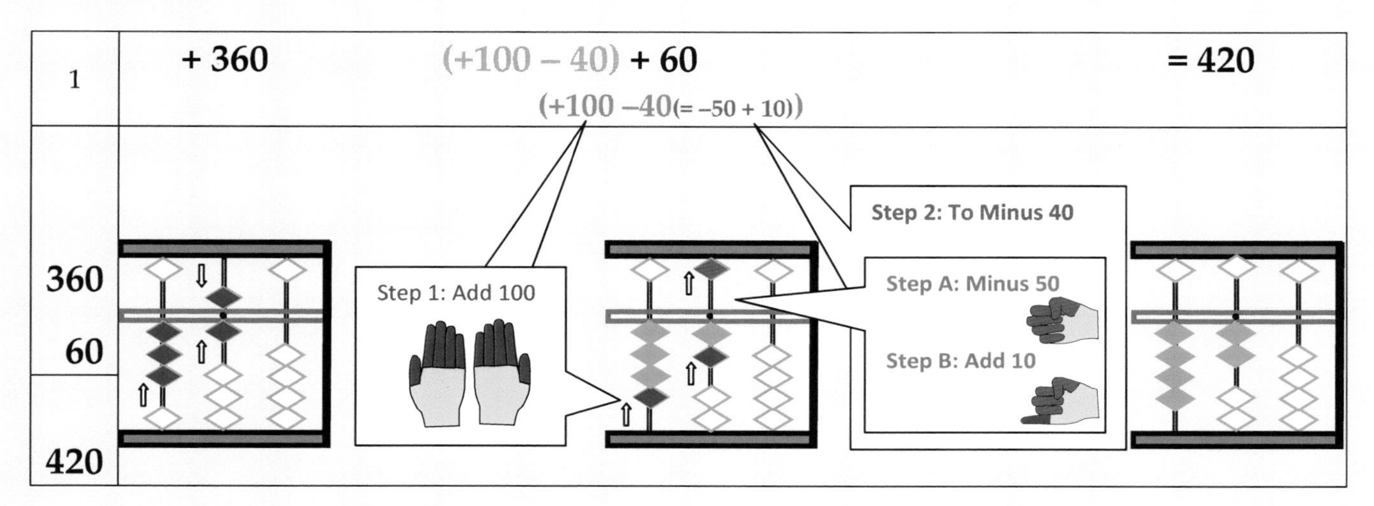

Problem	Action
+ 360	Move **three earth beads** to touch the beam on the **hundreds rod.** Move the **heaven bead and one earth bead** to touch the beam on the **tens rod.** There is nothing to do on the ones rod because ones place number is zero.
+ 60	Now we need to +60 on the tens rod, however we do not have enough earth beads to +60 on the tens rod. **When you want to +60 and you do not have enough beads:** **Use +60 = +100 – 40 Big Friend formula to do your calculations.** *(This can be taught as using the same bead movement as for +6, but add on the hundreds rod and minus on the tens rod.)* **Step 1: Add 100** – Move **one earth bead up** to touch the beam on the **hundreds rod.** ***Now we need to finish the formula by doing – 40 on the tens rod, but we do not have enough earth beads to do –40 so, GET HELP from small friends formula and do – 40.*** **Step 2: Minus 40** = **Step A: Minus 50** – Move the heaven bead up to touch the frame on the tens rod. **Step B: Add 10** – Move one earth bead up to touch the beam on the tens rod. There is nothing to do on the ones rod because the ones place number is zero.

ATTENTION

- Ask students to say the formula while they use it. This makes it easy for them to understand and follow through with all the steps in the formula.
- **Do not** say the small friend formula even when you are getting help from it.
- For instance, in example 4, JUST SAY "– 40" while performing the small friend formula bead movements. DO NOT say the Small friend formula out loud.

EXAMPLE: 5 + 60 = + 100 – 40

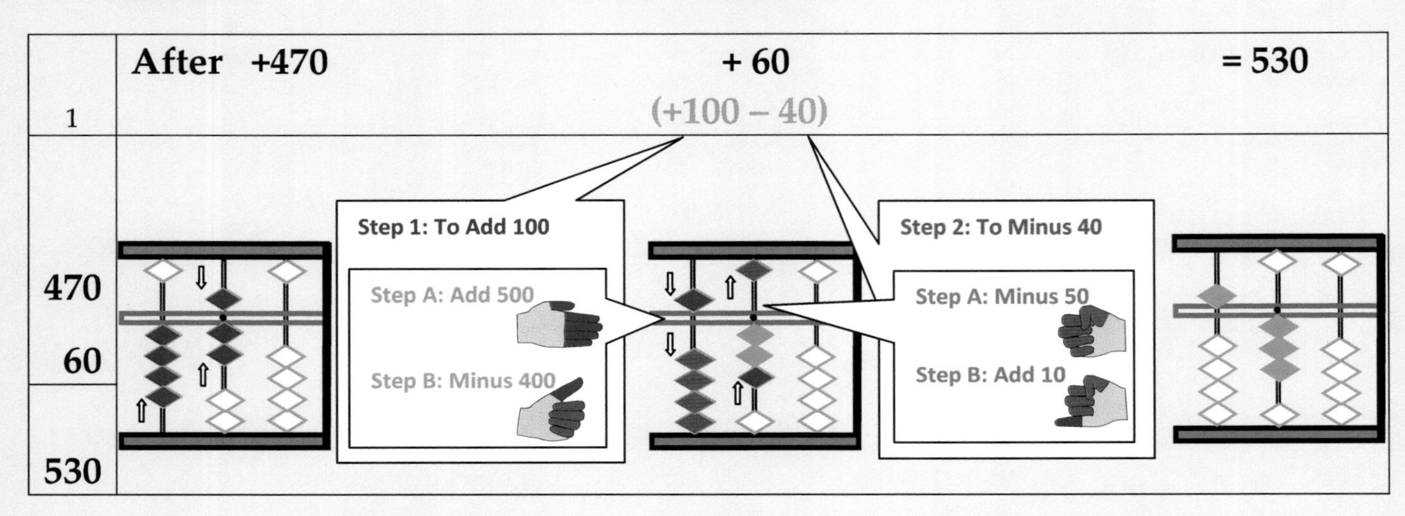

Problem	Action
+ 470	Move **four earth beads** to touch the beam on the **hundreds rod.** Move the **heaven bead** and **two earth beads** to touch the beam on the **tens rod**. There is nothing to do on the ones rod because ones place number is zero.
+ 60	Now we need to +60 on the tens rod, however we do not have enough beads on the tens rod to +60. So, now we need to make use of the fact that **60 + 40 = 100** **When you want to +60 and you do not have enough beads:** **Use +60 = +100 – 40 Big Friend formula to do your calculations.** ***Now to do +100 you do not have enough beads so, GET HELP from small friend formula and do +100.*** **Step 1: Add 100** = **Step A: Add 500** – Move the heaven bead down to touch the beam on the hundreds rod. **Step B: Minus 400** – Move all four earth beads down to touch the frame on the hundreds rod. ***Now we need to finish the formula by doing – 40 on the tens rod, but we do not have enough earth beads to do –40 so, GET HELP from small friend formula and do – 40.*** **Step 2: Minus 40** = **Step A: Minus 50** – Move the heaven bead up to touch the frame on the tens rod. **Step B: Add 10** – Move one earth bead up to touch the beam on the tens rod. ***(When we do + 100 and – 40, we get to keep 60 in our game.)*** There is nothing to do on the ones rod because the ones place number is zero

LESSON 9 – SAMPLE PROBLEMS

+ 6 = + 10 – 4 **+ 60 = + 100 – 40**

TO INTRODUCE –4 OR –40 BY GETTING HELP FROM SMALL FRIENDS

FORMULA OF –4 = –5 +1 OR –40 = –50 +10

Work these problems a few times to study and understand the concept and the relationship between the beads moved.

1	2	3	4	5	6	7	8	9	10
09	08	55	66	27	69	70	85	25	55
- 04	- 04	06	06	36	66	60	60	39	66
- 04	06	- 40	60	60	- 04	60	26	60	66

10:S:1

© SAI Speed Math Academy, USA

TO INTRODUCE +10 OR +100 BY GETTING HELP FROM SMALL FRIENDS

FORMULA OF +10 = +50 –40 OR +100 = +500 –400

Work these problems a few times to study and understand the concept and the relationship between the beads moved.

1	2	3	4	5	6	7	8	9	10
17	29	46	49	55	80	48	58	248	155
36	26	08	16	96	67	16	96	268	360
- 40	16	60	- 44	- 40	16	60	60	36	19

10:S:2

© SAI Speed Math Academy, USA

1	2	3	4	5	6	7	8	9	10
13	14	25	25	17	28	16	55	355	177
26	36	46	- 11	46	37	36	96	269	367
26	06	64	26	60	- 20	- 21	14	33	- 401
- 41	- 12	16	65	- 11	16	64	86	- 442	368

10:S:3

© SAI Speed Math Academy, USA

POINTS TO REMEMBER

The rows above consist of sample problems to introduce this week's formula. Explain to your child when and how to use the formula. Work with the sample problems until your child understands the formula and that the formulas are to be used ONLY when there are not enough beads to add or subtract.

WEEK 11 – LESSON 10 – INTRODUCING + 5 CONCEPT

LESSON 10 – EXAMPLE

CONCEPTS OF THE WEEK

TO ADD = ADD 10, LESS BIG FRIEND **+ 5 = + 10 – 5** **+ 50 = + 100 – 50**

EXAMPLE: 1

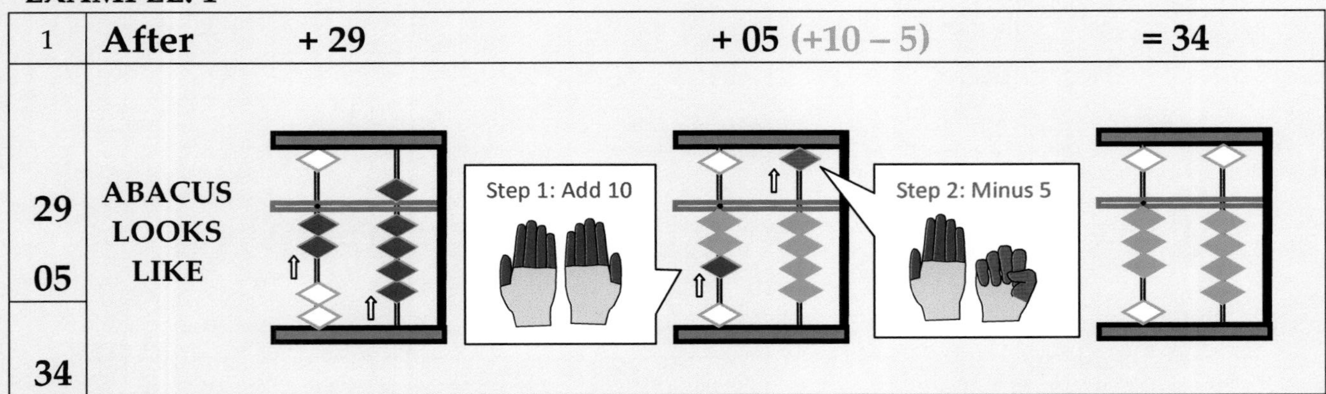

Problem	Action
+ 29	Move **two earth beads** to touch the beam on the **tens rod.** Move the **heaven bead and four earth beads** to touch the beam on the **ones rod.**
+ 05	There is nothing to do on the tens rod because tens place number is zero. Now we need to +5 on the ones rod, however we do not have enough beads on the ones rod to +5. (*The heaven bead is in the game, so we cannot get help from it.*) So, now we need to make use of the fact that **5 + 5 = 10** **When you want to +5 and you do not have enough beads:** **Use +5 = +10 – 5 Big Friend formula to do your calculations.** Step 1: Add 10 – Move **one earth bead up** to touch the beam on the **tens rod.** *(One earth bead on the tens rod is equal to '10')* *We know that there is a five in the ten (5 + 5 = 10), so let us get help from 10 by adding it to our game. However, we were supposed to +5, instead we did +10, which means we have 5 more than what we need. So, now we have to send the 5 away from our game.* Step 2: Minus 5 – Move **heaven bead away** from the beam on the **ones rod**. *(We move heaven bead out of the game because it is equal to five.)* ***(When we do + 10 and – 5, we get to keep 5 in our game.)*** **+10 – 5 = +5**

EXAMPLE: 2

1	After	+ 48	+ 05 (+10 (= +50 –40) – 5)	= 53
48 05 53	ABACUS LOOKS LIKE	Step 1: To Add 10 Step A: Add 50 Step B: Minus 40	Step 2: Minus 5	

Problem	Action
+ 48	Move **four earth beads** to touch the beam on the **tens rod.** Move the **heaven bead and three earth beads** to touch the beam on the **ones rod.**
+ 05	There is nothing to do on the tens rod because tens place number is zero. Now we need to +5 on the ones rod, however we do not have enough beads on the ones rod to +5. *(The heaven bead is in the game, so we cannot get help from it.)* So, now we need to make use of the fact that **5 + 5 = 10** **When you want to +5 and you do not have enough beads:** **Use +5 = +10 – 5 Big Friend formula to do your calculations.** ***Now to do +10 you do not have enough earth beads so, GET HELP from small friend formula and do +10.*** **Step 1: Add 10** = **Step A: Add 50** – Move the heaven bead down to touch the beam on the tens rod. **Step B: Minus 40** – Move all four earth beads away from the beam on the tens rod. **Now complete the +5 formula by doing – 5 on the ones rod.** **Step 2: Minus 5** – Move **heaven bead away** from the beam on the **ones rod**. *(We move heaven bead out of the game because it is equal to five.)* ***(When we do + 10 and – 5, we get to keep 5 in our game.)*** **+10 – 5 = +5**

ATTENTION

Sometimes to +10, students will have to get help from small friends formula of +10 = +50 – 40 and to do +100 they might have to use +100 = +500 – 400 (small friends formula for +10 on hundreds rod). Example: 95 + 55 = 150 or 495 +55 = 550
Please make sure that your student understands that they have to get help from +10 Small Friends formula and then finish by doing –5 or – 50 (of +5 or +50 formula) depending on which rod they are adding. Some students get confused with this concept in this lesson.

EXAMPLE: 3 **+ 50 = + 100 – 50**

1	**After** **+ 60** (+100 – 50) **+ 50** **= 110**
60 **50** **110**	**ABACUS LOOKS LIKE** Step 1: Add 100 Step 2: Minus 50 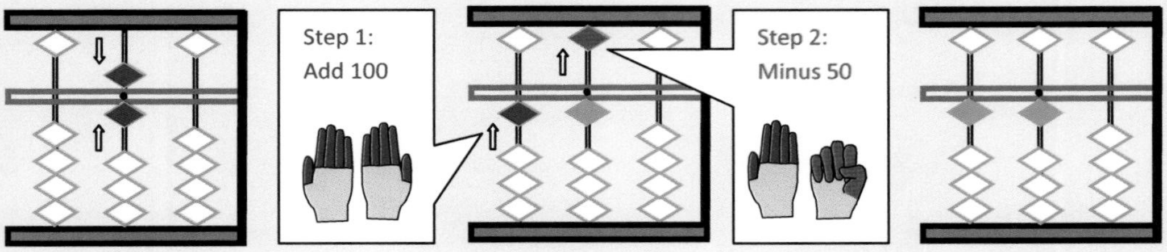

Problem	Action
+ 60	Move the **heaven bead** and **one earth bead** to touch the beam on the **tens rod.** There is nothing to do on the ones rod because ones place number is zero.
+ 50	Now we need to +50 on the tens rod, however we do not have enough earth beads to +50 on the tens rod. **When you want to +50 and you do not have enough beads:** **Use +50 = +100 – 50 Big Friend formula to do your calculations.** *(This can be taught as using the same bead movement as for +5, but add on the hundreds rod and minus on the tens rod.)* **Step 1: Add 100** – Move **one earth bead up** to touch the beam on the **hundreds rod**. **Step 2: Minus 50** – Move the **heaven bead away** from the beam on the **tens rod**. *(We move heaven bead on the tens rod out of the game because it equals to 50.)* There is nothing to do on the ones rod because ones place number is zero.

ATTENTION

- Ask students to say the formula while they use it. This makes it easy for them to understand and follow through with all the steps in the formula.
- When adding 15 to another number (where they need to use +5 formula) students will +10 once and then **–5**. Make sure they understand that they have to +10 once for the ten in the 15 and **another** +10 to use the formula before finishing with **–5** while +5. Eg: 39 + 15
- When following the formula on the tens rod, students usually +100 but then get confused and try to do –5 instead of –50. Make them understand that big friend of 50 is 50 and they help each other.

LESSON 10 – SAMPLE PROBLEMS

+ 5 = + 10 – 5 **+ 50 = + 100 – 50**

TO INTRODUCE +5 = + 10 – 5 and +50 = + 100 – 50 FORMULA

Work these problems a few times to study and understand the concept and the relationship between the beads moved.

1	2	3	4	5	6	7	8	9	10
		05	25	56	57	64	83	159	257
09	07	05	15	05	50	51	54	150	260
05	15	05	05	15	05	15	55	15	25

11:S:1

© SAI Speed Math Academy, USA

TO INTRODUCE +10 OR +100 BY GETTING HELP FROM SMALL FRIENDS FORMULA OF +10 = +50 –40 OR +100 = +500 –400

Work these problems a few times to study and understand the concept and the relationship between the beads moved.

1	2	3	4	5	6	7	8	9	10
		15	25	28	37	55	98	196	266
16	25	25	35	25	15	55	50	55	250
35	25	15	50	51	55	55	15	55	35

11:S:2

© SAI Speed Math Academy, USA

1	2	3	4	5	6	7	8	9	10
49	99	39	39	18	23	17	159	234	165
45	50	25	15	25	24	15	205	75	75
51	15	35	05	08	25	17	155	56	55
25	- 44	55	95	54	55	15	- 209	85	255

11:S:3

© SAI Speed Math Academy, USA

POINTS TO REMEMBER

The rows above consist of sample problems to introduce this week's formula. Explain to your child when and how to use the formula. Work with the sample problems until your child understands the formula and that the formulas are to be used ONLY when there are not enough beads to add or subtract.

WEEK 12 – LESSON 11 – INTRODUCING +4 CONCEPT

LESSON 11 – EXAMPLE

CONCEPTS OF THE WEEK

TO ADD = ADD 10, LESS BIG FRIEND   + 4 = + 10 – 6   + 40 = + 100 – 60

EXAMPLE: 1

1	After	+ 18	+ 04 (+10 – 6)	= 22
18 04 22	ABACUS LOOKS LIKE	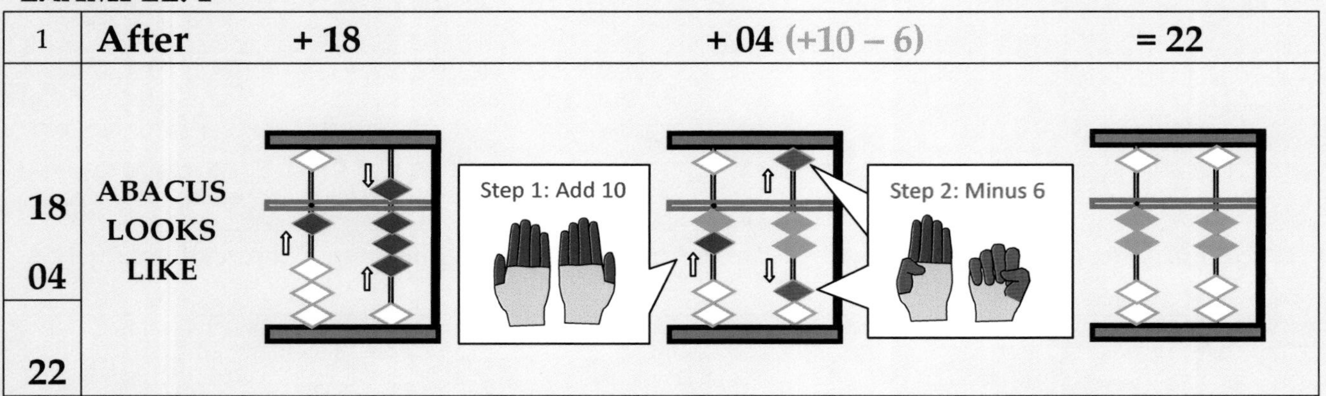		

Problem	Action
+ 18	Move **one earth bead** to touch the beam on the **tens rod.** Move he **heaven bead and three earth beads** to touch the beam on the **ones rod.**
+ 04	There is nothing to do on the tens rod because tens place number is zero. Now we need to +4 on the ones rod, however we do not have enough beads on the ones rod to +4. *(The heaven bead is in the game, so we cannot get help from it.)* So, now we need to make use of the fact that **4 + 6 = 10** **When you want to +4 and you do not have enough beads:** **Use +4 = +10 – 6 Big Friend formula to do your calculations.** **Step 1: Add 10** – Move **one earth bead up** to touch the beam on the **tens rod**. *(One earth bead on the tens rod is equal to '10')* *We know that there is a four in the ten (4 + 6 = 10), so let us get help from 10 by adding it to our game. However, we were supposed to +4, instead we did +10, which means we have 6 more than what we need. So, now we have to send the 6 away from our game.* **Step 2: Minus 6** – Move **heaven bead and one earth bead away** from the beam on the **ones rod**. ***(When we do + 10 and – 6, we get to keep 4 in our game.)*** **+10 – 6 = +4**

EXAMPLE: 2

1	After	+ 47	+ 04 (+10 (= +50 –40) – 6)	= 51
47 04 51	ABACUS LOOKS LIKE	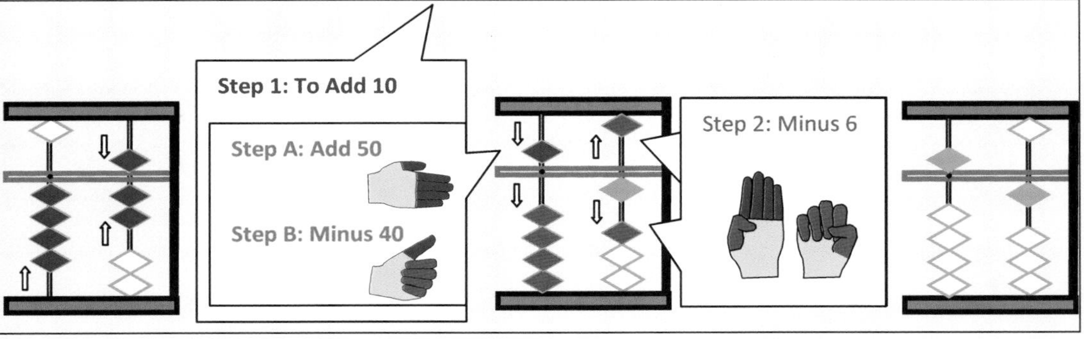	Step 1: To Add 10 Step A: Add 50 Step B: Minus 40 Step 2: Minus 6	

Problem	Action
+ 47	Move **four earth beads** to touch the beam on the **tens rod.** Move the **heaven bead and two earth beads** to touch the beam on the **ones rod.**
+ 04	There is nothing to do on the tens rod because tens place number is zero. Now we need to +4 on the ones rod, however we do not have enough beads on the ones rod to +4. (*The heaven bead is in the game, so we cannot get help from it.)* So, now we need to make use of the fact that **4 + 6 = 10** **When you want to +4 and you do not have enough beads:** **Use +4 = +10 – 6 Big Friend formula to do your calculations.** ***Now to do +10 you do not have enough earth beads so, GET HELP from small friend formula and do +10.*** **Step 1: Add 10** = Step A: Add 50 – Move the heaven bead down to touch the beam on the tens rod. Step B: Minus 40 – Move all four earth beads away from the beam on the tens rod. **Now complete the +4 formula by doing – 6 on the ones rod.** **Step 2: Minus 6** – Move the **heaven bead and one earth bead away** from the beam on the **ones rod**. ***(When we do + 10 and – 6, we get to keep 4 in our game.)*** **+10 – 6 = +4**

ATTENTION

Ask students to say the formula while they use it. This makes it easy for them to understand and follow through with all the steps. This will also ensure that they use the correct formula. Some of them will confuse with small friend formula and big friend formula by doing +10 and then –1 instead of doing – 6.

EXAMPLE: 3 **+ 40 = + 100 – 60**

1	After + 90 (+100 – 60) + 40 = 130
90 40 130	ABACUS LOOKS LIKE 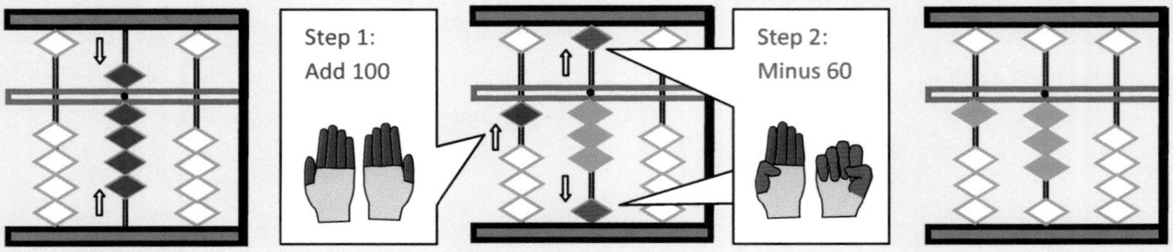

Problem	Action
+ 90	Move the **heaven bead** and **four earth beads** to touch the beam on the **tens rod.** There is nothing to do on the ones rod because ones place number is zero.
+ 40	Now we need to +40 on the tens rod, however we do not have enough earth beads to +40 on the tens rod. **When you want to +40 and you do not have enough beads:** **Use +40 = +100 – 60 Big Friend formula to do your calculations.** *(This can be taught as using the same bead movement as for +4, but add on the hundreds rod and minus on the tens rod.)* **Step 1: Add 100** – Move **one earth bead up** to touch the beam on the **hundreds rod**. **Step 2: Minus 60** – Move the **heaven bead and one earth bead away** from the beam on the **tens rod**. There is nothing to do on the ones rod because ones place number is zero.

ATTENTION

From this week we will learn to add 4, 3, 2, and 1 using Big Friends formula. Please make sure that children understand that 4, 3, 2, and 1 have small friends formula and big friends formula.

- When the heaven bead is not in the game then they have to use small friends formula (+4 = +5 – 1) to finish calculating.
- When the heaven bead is in the game AND there are not enough earth beads to +4 then they have to use Big Friends formula (+4 = +10 – 6).
- Some children will try to add the heaven bead and then will try to follow the big friend formula. Care has to be taken to make them understand and analyze which formula they should use.

LESSON 11 – SAMPLE PROBLEMS

+ 4 = + 10 – 6 **+ 40 = + 100 – 60**

TO INTRODUCE +4 = + 10 – 6 and +40 = + 100 – 60 FORMULA

Work these problems a few times to study and understand the concept and the relationship between the beads moved.

1	2	3	4	5	6	7	8	9	10	12:S:1
04	03	06	15	20	50	57	72	140	157	
04	04	04	04	40	40	04	44	144	140	
04	14	04	04	40	40	44	44	44	44	

© SAI Speed Math Academy, USA

TO INTRODUCE +10 OR +100 BY GETTING HELP FROM SMALL FRIENDS FORMULA OF +10 = +50 –40 OR +100 = +500 –400

Work these problems a few times to study and understand the concept and the relationship between the beads moved.

1	2	3	4	5	6	7	8	9	10	12:S:2
		49	25	48	37	66	94			
17	16	04	24	14	14	44	44	196	266	
44	34	14	04	40	44	44	14	344	144	

© SAI Speed Math Academy, USA

1	2	3	4	5	6	7	8	9	10	12:S:3
24	56	29	97	59	28	29	27	148	178	
44	- 13	34	47	84	44	14	25	44	76	
05	29	15	08	43	13	18	28	- 131	11	
42	44	45	94	68	- 61	50	45	44	48	

© SAI Speed Math Academy, USA

POINTS TO REMEMBER

The rows above consist of sample problems to introduce this week's formula. Explain to your child when and how to use the formula. Work with the sample problems until your child understands the formula and that the formulas are to be used ONLY when there are not enough beads to add or subtract.

WEEK 13 – LESSON 12 – INTRODUCING +3 CONCEPT

LESSON 12 – EXAMPLE

CONCEPTS OF THE WEEK

TO ADD = ADD 10, LESS BIG FRIEND **+ 3 = + 10 – 7** **+ 30 = + 100 – 70**

EXAMPLE: 1

1	After	+ 27	+ 03 (+10 – 7)	= 30
27 03 30	ABACUS LOOKS LIKE	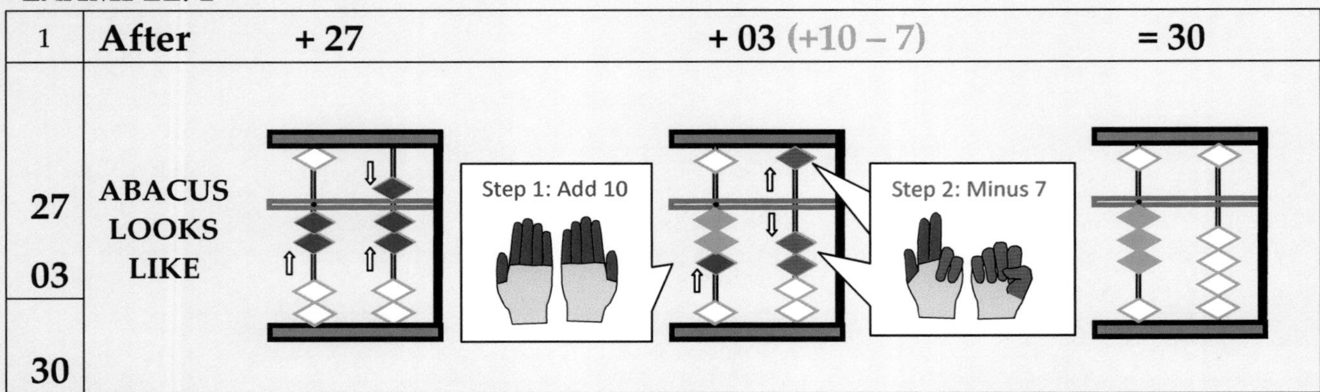		

Problem	Action
+ 27	Move **two earth beads** to touch the beam on the **tens rod.** Move the **heaven bead and two earth beads** to touch the beam on the **ones rod.**
+ 03	There is nothing to do on the tens rod because tens place number is zero. Now we need to +3 on the ones rod, however we do not have enough beads on the ones rod to +3. *(The heaven bead is in the game, so we cannot get help from it.)* So, now we need to make use of the fact that **3 + 7 = 10** **When you want to +3 and you do not have enough beads:** **Use +3 = +10 – 7 Big Friend formula to do your calculations.** Step 1: Add 10 – Move **one earth bead up** to touch the beam on the **tens rod**. *(One earth bead on the tens rod is equal to '10')* *We know that there is a three in the ten (3 + 7 = 10), so let us get help from 10 by adding it to our game. However, we were supposed to +3, instead we did +10, which means we have 7 more than what we need. So, now we have to send the 7 away from our game.* Step 2: Minus 7 – Move the **heaven bead and two earth beads away** from the beam on the **ones rod**. ***(When we do + 10 and – 7, we get to keep 3 in our game.)*** **+10 – 7 = +3**

EXAMPLE: 2

1	After	+ 49	+ 03 (+10 (= +50 –40) – 7)	= 52
49 03 52	ABACUS LOOKS LIKE		Step 1: To Add 10 Step A: Add 50 Step B: Minus 40 Step 2: Minus 7	

Problem	Action
+ 49	Move **four earth beads** to touch the beam on the **tens rod.** Move the **heaven bead and four earth beads** to touch the beam on the **ones rod.**
+ 03	There is nothing to do on the tens rod because tens place number is zero. Now we need to +3 on the ones rod, however we do not have enough beads on the ones rod to +3. (*The heaven bead is in the game, so we cannot get help from it.*) So, now we need to make use of the fact that **3 + 7 = 10** **When you want to +3 and you do not have enough beads:** **Use +3 = +10 – 7 Big Friend formula to do your calculations.** ***Now to do +10 you do not have enough earth beads so, GET HELP from small friend formula and do +10.*** **Step 1: Add 10** = **Step A: Add 50** – Move the heaven bead down to touch the beam on the tens rod. **Step B: Minus 40** – Move all four earth beads away from the beam on the tens rod. **Now complete the +3 formula by doing – 7 on the ones rod.** **Step 2: Minus 7** – Move **heaven bead and two earth beads away** from the beam on the **ones rod**. ***(When we do + 10 and – 7, we get to keep 3 in our game.)*** **+10 – 7 = +3**

ATTENTION

Ask students to say the formula while they use it. This makes it easy for them to understand and follow through with all the steps. This will also ensure that they use the correct formula. Some of them will confuse with small friend formula and big friend formula by doing +10 and then –2 instead of doing – 7.

EXAMPLE: 3 $+30 = +100 - 70$

1	After	+ 80	(+100 – 70) + 30	= 110
80 30 110	ABACUS LOOKS LIKE	Step 1: Add 100	Step 2: Minus 70	

Problem	Action
+ 80	Move the **heaven bead** and **three earth beads** to touch the beam on the **tens rod.** There is nothing to do on the ones rod because ones place number is zero.
+ 30	Now we need to +30 on the tens rod, however we do not have enough earth beads to +30 on the tens rod. **When you want to +30 and you do not have enough beads:** **Use +30 = +100 – 70 Big Friend formula to do your calculations.** *(This can be taught as using the same bead movement as for +3, but add on the hundreds rod and minus on the tens rod.)* **Step 1: Add 100** – Move **one earth bead up** to touch the beam on the **hundreds rod**. **Step 2: Minus 70** – Move the **heaven bead and two earth beads away** from the beam on the **tens rod**. There is nothing to do on the ones rod because ones place number is zero.

ATTENTION

Please make sure that children understand that 4, 3, 2, and 1 have a small friend formula and a big friend formula.

- When the heaven bead is not in the game then they have to use small friends formula (+3 = +5 – 2) to finish calculating.
- When the heaven bead is in the game AND there are not enough earth beads to +3 then they have to use Big Friends formula (+3 = +10 – 7).
- Some children will try to add the heaven bead and then will try to follow the big friend formula. Care has to be taken to make them understand and analyze which formula they should use.

LESSON 12 – SAMPLE PROBLEMS

+ 3 = + 10 – 7 **+ 30 = + 100 – 70**

TO INTRODUCE +3 = + 10 – 7 and +30 = + 100 – 70 FORMULA

Work these problems a few times to study and understand the concept and the relationship between the beads moved.

1	2	3	4	5	6	7	8	9	10	
03	05	29	12	60	34	38	72	79	185	13:S:1
03	03	33	73	30	33	33	33	13	33	
03	03	03	30	30	03	30	- 03	33	53	

© SAI Speed Math Academy, USA

TO INTRODUCE +10 OR +100 BY GETTING HELP FROM SMALL FRIENDS

FORMULA OF +10 = +50 –40 OR +100 = +500 –400

Work these problems a few times to study and understand the concept and the relationship between the beads moved.

1	2	3	4	5	6	7	8	9	10	
		48	24	39	88	97	49			13:S:2
38	19	03	63	13	30	31	13	289	279	
33	33	35	33	53	33	13	33	333	233	

© SAI Speed Math Academy, USA

1	2	3	4	5	6	7	8	9	10	
95	55	65	81	75	67	89	51	194	179	13:S:3
46	19	66	32	47	82	36	62	31	134	
08	33	55	52	33	13	25	75	29	62	
13	43	- 43	- 41	33	- 22	- 30	63	- 13	78	

© SAI Speed Math Academy, USA

POINTS TO REMEMBER

The rows above consist of sample problems to introduce this week's formula. Explain to your child when and how to use the formula. Work with the sample problems until your child understands the formula and that the formulas are to be used ONLY when there are not enough beads to add or subtract.

WEEK 14 – LESSON 13 – INTRODUCING +2 CONCEPT

LESSON 13 – EXAMPLE

CONCEPTS OF THE WEEK

TO ADD = ADD 10, LESS BIG FRIEND **+ 2 = + 10 – 8** **+ 20 = + 100 – 80**

EXAMPLE: 1

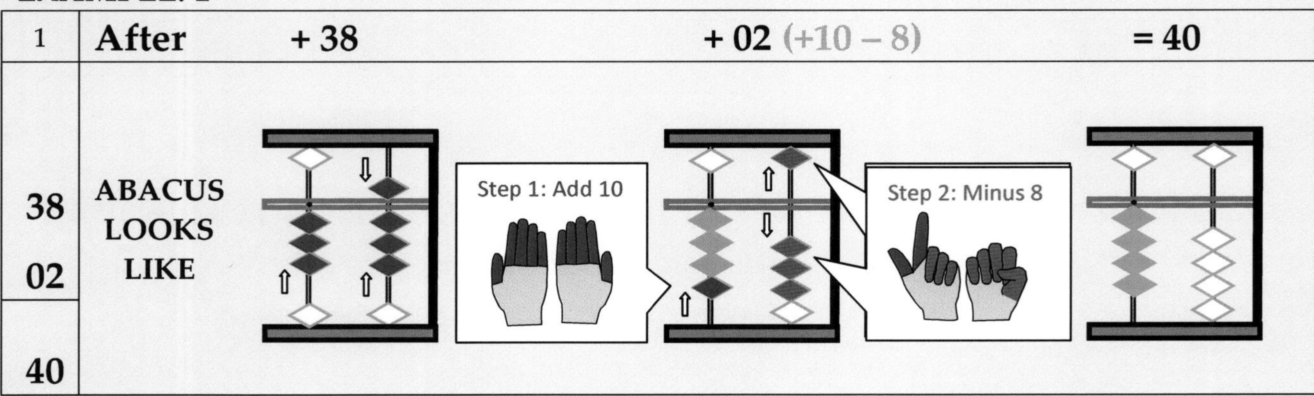

Problem	Action
+ 38	Move **three earth beads** to touch the beam on the **tens rod.** Move the **heaven bead and three earth beads** to touch the beam on the **ones rod.**
+ 02	There is nothing to do on the tens rod because tens place number is zero. Now we need to +2 on the ones rod, however we do not have enough beads on the ones rod to +2. (*The heaven bead is in the game, so we cannot get help from it.*) So, now we need to make use of the fact that **2 + 8 = 10** **When you want to +2 and you do not have enough beads:** **Use +2 = +10 – 8 Big Friend formula to do your calculations.** **Step 1: Add 10** – Move **one earth bead up** to touch the beam on the **tens rod**. *(One earth bead on the tens rod is equal to '10')* *We know that there is a two in the ten (2 + 8 = 10), so let us get help from 10 by adding it to our game. However, we were supposed to +2, instead we did +10, which means we have 8 more than what we need. So, now we have to send the 8 away from our game.* **Step 2: Minus 8** – Move the **heaven bead and three earth beads away** from the beam on the **ones rod**. ***(When we do + 10 and – 8, we get to keep 2 in our game.)*** **+10 – 8 = +2**

EXAMPLE: 2

1	After	+ 49	+ 02 (+10 (= +50 –40) – 8)	= 51
49 02 51	ABACUS LOOKS LIKE		Step 1: To Add 10 Step A: Add 50 Step B: Minus 40	Step 2: Minus 8

Problem	Action
+ 49	Move **four earth beads** to touch the beam on the **tens rod.** Move the **heaven bead and four earth beads** to touch the beam on the **ones rod.**
+ 02	There is nothing to do on the tens rod because tens place number is zero. Now we need to +2 on the ones rod, however we do not have enough beads on the ones rod to +2. *(The heaven bead is in the game, so we cannot get help from it.)* So, now we need to make use of the fact that **2 + 8 = 10** **When you want to +2 and you do not have enough beads:** **Use +2 = +10 – 8 Big Friend formula to do your calculations.** ***Now to do +10 you do not have enough earth beads so, GET HELP from small friend formula and do +10.*** **Step 1: Add 10** = **Step A: Add 50** – Move the heaven bead down to touch the beam on the tens rod. **Step B: Minus 40** – Move all four earth beads away from the beam on the tens rod. **Now complete the +2 formula by doing – 8 on the ones rod.** **Step 2: Minus 8** – Move the **heaven bead and three earth beads away** from the beam on the **ones rod**. ***(When we do + 10 and – 8, we get to keep 2 in our game.)*** **+10 – 8 = +2**

ATTENTION

Ask students to say the formula while they use it. This makes it easy for them to understand and follow through with all the steps. This will also ensure that they use the correct formula. Some of them will confuse with small friend formula and big friend formula by doing +10 and then –3 instead of doing – 8.

EXAMPLE: 3 **+ 20 = + 100 – 80**

1	**After** **+ 80** (+100 – 80) **+ 20** **= 100**
80 20 100	**ABACUS LOOKS LIKE** 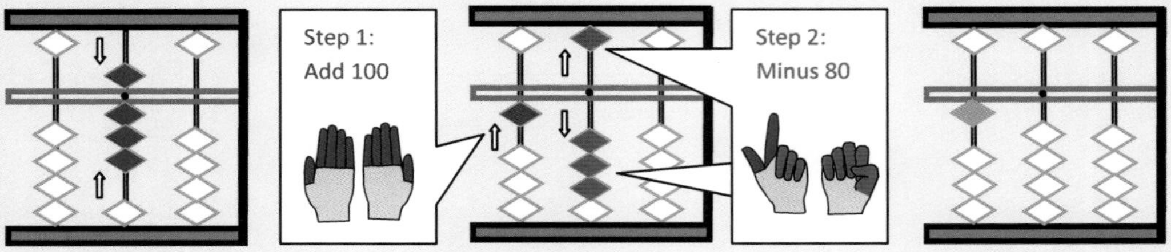 Step 1: Add 100 Step 2: Minus 80

Problem	Action
+ 80	Move the **heaven bead** and **three earth beads** to touch the beam on the **tens rod.** There is nothing to do on the ones rod because ones place number is zero.
+ 20	Now we need to +20 on the tens rod, however we do not have enough earth beads to +20 on the tens rod. **When you want to +20 and you do not have enough beads:** **Use +20 = +100 – 80 Big Friend formula to do your calculations.** *(This can be taught as using the same bead movement as for +2, but add on the hundreds rod and minus on the tens rod.)* **Step 1: Add 100** – Move **one earth bead up** to touch the beam on the **hundreds rod**. **Step 2: Minus 80** – Move the **heaven bead and three earth beads away** from the beam on the **tens rod.** There is nothing to do on the ones rod because ones place number is zero.

ATTENTION

Please make sure that children understand that 4, 3, 2, and 1 have a small friend formula and a big friend formula.

- When the heaven bead is not in the game then they have to use small friends formula (+2 = +5 – 3) to finish calculating.
- When the heaven bead is in the game AND there are not enough earth beads to +2 then they have to use Big Friends formula (+2 = +10 – 8).
- Some children will try to add the heaven bead and then will try to follow the big friend formula. Care has to be taken to make them understand and analyze which formula they should use.

LESSON 13 – SAMPLE PROBLEMS

+ 2 = + 10 – 8 **+ 20 = + 100 – 80**

TO INTRODUCE +2 = + 10 – 8 and +20 = + 100 – 80 FORMULA

Work these problems a few times to study and understand the concept and the relationship between the beads moved.

1	2	3	4	5	6	7	8	9	10	
04	28	29	47	40	94	38	99	229	187	14:S:1
02	02	12	22	40	22	22	20	112	222	
02	20	24	12	20	22	- 60	02	225	12	

© SAI Speed Math Academy, USA

TO INTRODUCE +10 OR +100 BY GETTING HELP FROM SMALL FRIENDS FORMULA OF +10 = +50 –40 OR +100 = +500 –400

Work these problems a few times to study and understand the concept and the relationship between the beads moved.

1	2	3	4	5	6	7	8	9	10	
		48	54	39	98	97	99			14:S:2
49	28	02	94	12	22	22	50	158	498	
22	22	- 20	12	- 20	80	32	12	392	22	

© SAI Speed Math Academy, USA

1	2	3	4	5	6	7	8	9	10	
49	95	42	63	55	74	82	97	164	159	14:S:3
47	- 41	24	24	22	25	27	22	376	192	
22	82	86	39	32	22	42	30	- 140	35	
28	63	44	26	44	33	16	12	95	126	
17	22	24	- 41	- 31	91	85	- 40	29	- 411	

© SAI Speed Math Academy, USA

POINTS TO REMEMBER

The rows above consist of sample problems to introduce this week's formula. Explain to your child when and how to use the formula. Work with the sample problems until your child understands the formula and that the formulas are to be used ONLY when there are not enough beads to add or subtract.

WEEK 15 – LESSON 14 – INTRODUCING +1 CONCEPT

This is a very important formula.

During Week 16 – Lesson 15 students will have to get help from this formula when used on the tens rod (+10 = +100 – 90) to complete all the big friends formula in specific situations. Example 99 + 02 = 101

So, please make sure that children are very clear in their understanding of this formula.

LESSON 14 – EXAMPLE

CONCEPTS OF THE WEEK

TO ADD = ADD 10, LESS BIG FRIEND   + 1 = + 10 – 9   + 10 = + 100 – 90

EXAMPLE: 1

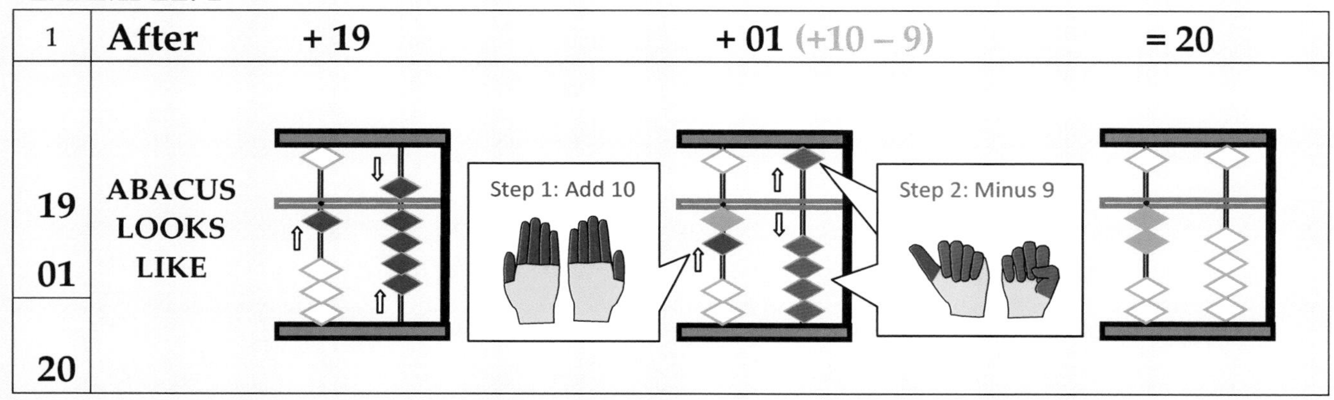

Problem	Action
+ 19	Move **one earth bead** to touch the beam on the **tens rod.** Move the **heaven bead and four earth beads** to touch the beam on the **ones rod.**
+ 01	There is nothing to do on the tens rod because tens place number is zero. Now we need to +1 on the ones rod, however we do not have enough beads on the ones rod to +1. *(The heaven bead is in the game, so we cannot get help from it.)* So, now we need to make use of the fact that **1 + 9 = 10** **When you want to +1 and you do not have enough beads:** **Use +1 = +10 – 9 Big Friend formula to do your calculations.** Step 1: Add 10 – Move **one earth bead up** to touch the beam on the **tens rod**. *(One earth bead on the tens rod is equal to '10')* *We know that there is a one in the ten (1 + 9 = 10), so let us get help from 10 by adding it to our game. However, we were supposed to +1, instead we did +10, which means we have 9 more than what we need. So, now we have to send the 9 away from our game.* Step 2: Minus 9 – Move the **heaven bead and four earth beads away** from the beam on the **ones rod**. ***(When we do + 10 and – 9, we get to keep 1 in our game.)*** **+10 – 9 = +1**

EXAMPLE: 2

1	After	+ 49	+ 01 (+10 (= +50 –40) – 9)	= 50
49 01 50	ABACUS LOOKS LIKE	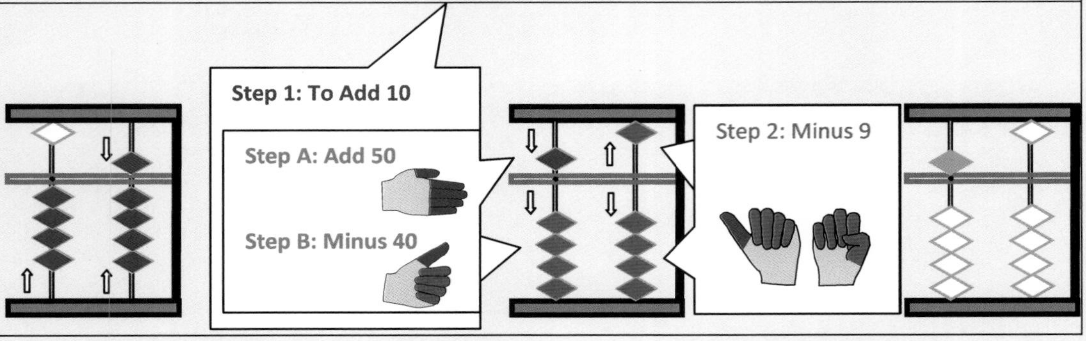		

Step 1: To Add 10

Step A: Add 50

Step B: Minus 40

Step 2: Minus 9

Problem	Action
+ 49	Move **four earth beads** to touch the beam on the **tens rod.** Move the **heaven bead and four earth beads** to touch the beam on the **ones rod.**
+ 01	There is nothing to do on the tens rod because tens place number is zero. Now we need to +1 on the ones rod, however we do not have enough beads on the ones rod to +1. *(The heaven bead is in the game, so we cannot get help from it.)* So, now we need to make use of the fact that **1 + 9 = 10** **When you want to +1 and you do not have enough beads:** **Use +1 = +10 – 9 Big Friend formula to do your calculations.** *Now to do +10 you do not have enough earth beads so, GET HELP from small friend formula and do +10.* **Step 1: Add 10** = Step A: Add 50 – Move the heaven bead down to touch the beam on the tens rod. Step B: Minus 40 – Move all four earth beads away from the beam on the tens rod. **Now complete the +1 formula by doing – 9 on the ones rod.** Step 2: Minus 9 – Move the **heaven bead and four earth beads away** from the beam on the **ones rod.** ***(When we do + 10 and – 9, we get to keep 1 in our game.)*** **+10 – 9 = +1**

ATTENTION

Ask students to say the formula while they use it. This makes it easy for them to understand and follow through with all the steps. This will also ensure that they use the correct formula. Some of them will confuse with small friend formula and big friend formula by doing +10 and then –4 instead of doing – 9.

EXAMPLE: 3 **+ 10 = + 100 – 90**

1	**After** **+ 90** (+100 – 90) **+ 10** **= 100**
90 10	ABACUS LOOKS LIKE 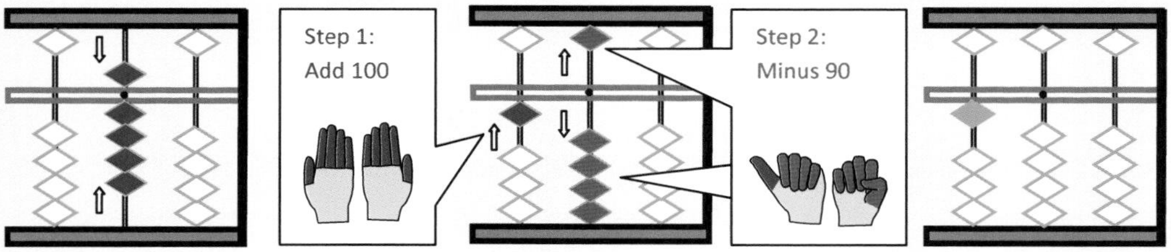 Step 1: Add 100 Step 2: Minus 90
100	

Problem	Action
+ 90	Move the **heaven bead** and **four earth beads** to touch the beam on the **tens rod.** There is nothing to do on the ones rod because ones place number is zero.
+ 10	Now we need to +10 on the tens rod, however we do not have enough earth beads to +10 on the tens rod. **When you want to +10 and you do not have enough beads:** **Use +10 = +100 – 90 Big Friend formula to do your calculations.** *(This can be taught as using the same bead movement as for +1, but add on the hundreds rod and minus on the tens rod.)* **Step 1: Add 100** – Move **one earth bead up** to touch the beam on the **hundreds rod**. **Step 2: Minus 90** – Move the **heaven bead and four earth beads away** from the beam on the **tens rod**. There is nothing to do on the ones rod because ones place number is zero.

ATTENTION

Please make sure that children understand that 4, 3, 2, and 1 have a small friend formula and a big friend formula.

- When the heaven bead is not in the game then they have to use small friends formula (+1 = +5 – 4) to finish calculating.
- When the heaven bead is in the game AND there are not enough earth beads to +1 then they have to use Big Friends formula (+1 = +10 – 9).
- Some children will try to add the heaven bead and then will try to follow the big friend formula. Care has to be taken to make them understand and analyze which formula they should use.

LESSON 14 – SAMPLE PROBLEMS

+ 1 = + 10 – 9   **+ 10 = + 100 – 90**

TO INTRODUCE +1 = + 10 – 9 and +10 = + 100 – 90 FORMULA

Work these problems a few times to study and understand the concept and the relationship between the beads moved.

1	2	3	4	5	6	7	8	9	10
44	49	29	49	55	94	95	99	194	144
11	11	11	41	39	11	18	10	111	151
- 23	- 20	15	10	11	- 01	41	31	- 201	111

15:S:1

© SAI Speed Math Academy, USA

TO INTRODUCE +10 OR +100 BY GETTING HELP FROM SMALL FRIENDS
FORMULA OF +10 = +50 –40 OR +100 = +500 –400

Work these problems a few times to study and understand the concept and the relationship between the beads moved.

1	2	3	4	5	6	7	8	9	10
			39	28	94	59	98	294	388
49	29	59	11	21	15	91	11	115	11
01	21	61	- 40	11	41	- 30	11	41	111

15:S:2

© SAI Speed Math Academy, USA

1	2	3	4	5	6	7	8	9	10
39	55	44	18	19	19	19	36	397	899
11	44	23	24	46	47	11	44	14	11
28	11	23	43	26	25	14	05	248	- 700
15	44	16	09	38	16	15	94	01	79
38	55	15	14	01	- 03	11	01	- 320	21

15:S:3

© SAI Speed Math Academy, USA

POINTS TO REMEMBER

The rows above consist of sample problems to introduce this week's formula. Explain to your child when and how to use the formula. Work with the sample problems until your child understands the formula and that the formulas are to be used ONLY when there are not enough beads to add or subtract.

WEEK 16 – LESSON 15 – INTRODUCING REGROUPING OR CARRY-OVER FROM ONES ROD TO HUNDREDS ROD

CONCEPTS OF THE WEEK – CARRYOVER TO THE HUNDRED'S ROD

TO ADD = ADD 10, LESS BIG FRIEND

TO ADD 10 = If you do not have enough beads to +10 in a formula

USE = +10 = +100 – 90 and/or +100 = +500 – 400

+9 = +10 – 1	**+4 = +10 – 6**
+8 = +10 – 2	**+3 = +10 – 7**
+7 = +10 – 3	**+2 = +10 – 8**
+6 = +10 – 4	**+1 = +10 – 9**
+5 = +10 – 5	**+10 = +100 – 90**

What do you need to succeed?

Passion, patience, practice, tenacity and commitment.

LESSON 15 – EXAMPLE

EXAMPLE 1: 99 + 01 = 100 (+1= +10 – 9)

In the example it is not possible to do +10 because the tens rod is full. So, we have to substitute **+10 = +100 – 90** and finish with the rest of the formula by doing minus 9 on the ones rod.

To do **+01 = +10** (+100 – 90) **– 9**

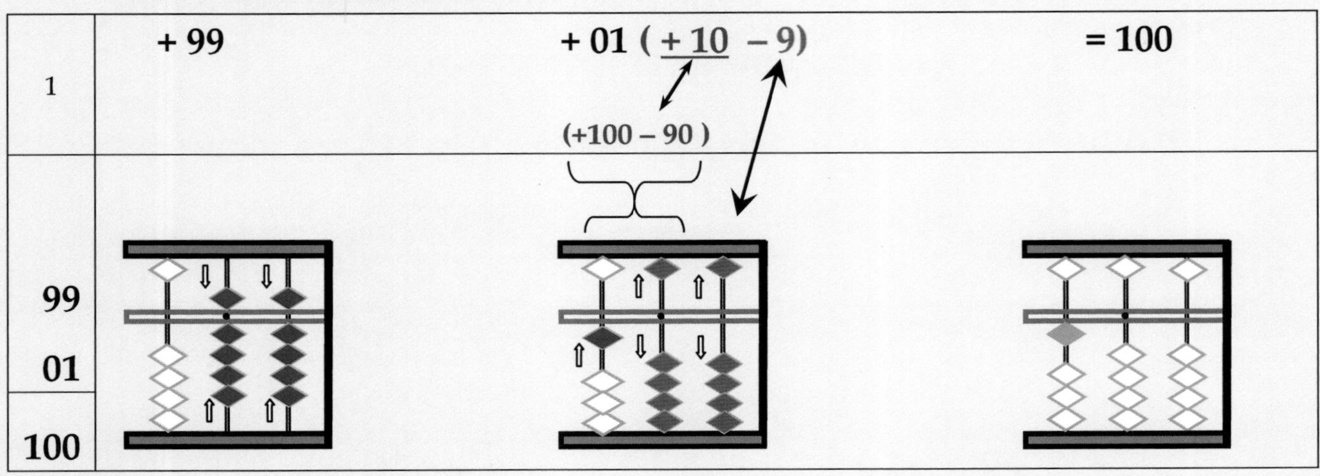

Problem	Action
+ 99	Move the **heaven bead and four earth beads** to touch the beam on the **tens rod.** Move the **heaven bead and four earth beads** to touch the beam on the **ones rod.**
+ 01	There is nothing to do on the tens rod because tens place number is zero. Now we need to **+1** on the ones rod, however we do not have enough beads on the ones rod to **+1**. So, now we need to use the big friend formula of **+1 = +10 – 9** ***Here, all the beads on the tens place rod are in the game, so we cannot do +10 directly. Now to do +10, you GET HELP from BIG FRIEND FORMULA of +10 and complete the step.*** *Big friend formula for +10 = +100 – 90* **We have to substitute +10 in the +1 = +10 – 9 with +10 = +100 – 90** **Now, +1 = +10** *(+100 – 90)* **– 9** **Step 1: Add 10 – Move one earth bead up on the hundreds rod** *(One earth bead on the hundreds rod is equal to '100')* **and do – 90 by moving the heaven bead and all earth beads away from the beam on the tens rod.** **Step 2: Minus 9 – Move the heaven bead and four earth beads away from the beam on the ones rod.** ***(Be sure to make students understand that – 9 has to be done on the ones rod because we are trying to +1 on the ones rod.)***

EXAMPLE 2: 96 + 07 = 103 (+7 = +10 – 3)

To do **+ 07 = +10 (**+100 – 90**) – 3**(–5 +2)

1	**+ 96**	**+ 07 (+ 10 – 3)** **(+100 – 90) (–5 +2)**	**= 103**
96 **07** **103**	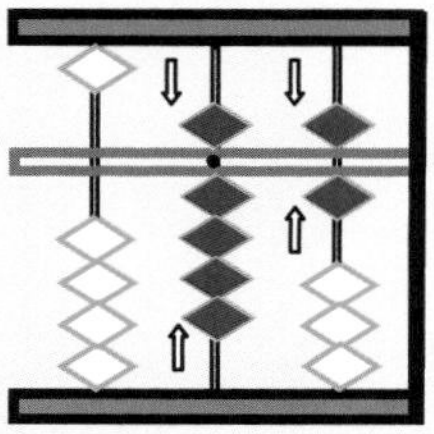	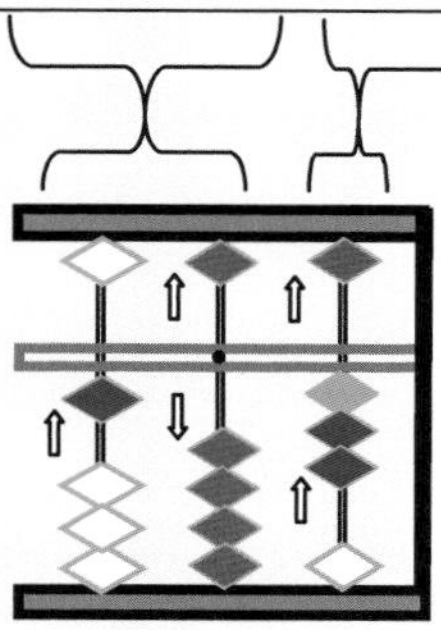	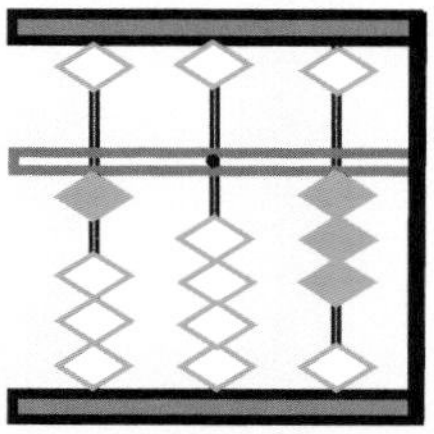

Problem	Action
+ 96	Move the **heaven bead and four earth beads** to touch the beam on the **tens rod.** Move the **heaven bead and one earth bead** to touch the beam on the **ones rod.**
+ 07	There is nothing to do on the tens rod because tens place number is zero. Now we need to +7 on the ones rod, however we do not have enough beads on the ones rod to +7. So, now we need to use the big friend formula of **+7 = +10 – 3** ***Here, all the beads on the tens place rod are in the game, so we cannot do +10 directly. Now to do +10, you GET HELP from BIG FRIEND FORMULA of +10 and complete the step.*** *Big friend formula for* **+10 = +100 – 90** **We have to substitute +10 in the +7 = +10 – 3 with +10 = +100 – 90** **Now, +7 = +10** *(+100 – 90)* **– 3** Step 1: Add 10 – **Move one earth bead up on the hundreds rod** *(One earth bead on the hundreds rod is equal to '100')* **and do – 90 by moving the heaven bead and all earth beads away from the beam on the tens rod.** Step 2: Minus 3 = *GET HELP from SMALL FRIENDS FORMULA OF – 3.* – **Move the heaven bead up to touch the frame on the ones rod and move two earth beads up to touch the beam on the ones rod.** ***(Be sure to make students understand that – 3 has to be done on the ones rod because we are trying to +7 on the ones rod.)***

EXAMPLES 3: 499 + 05 = 504 (+5 = +10 – 5)

Formula for **+5 = +10 – 5**

At times you may not have enough beads to use the **+10 = +100 – 90** formula.

In such case you may have to use **+10 = +100 (+500 – 400) – 90**

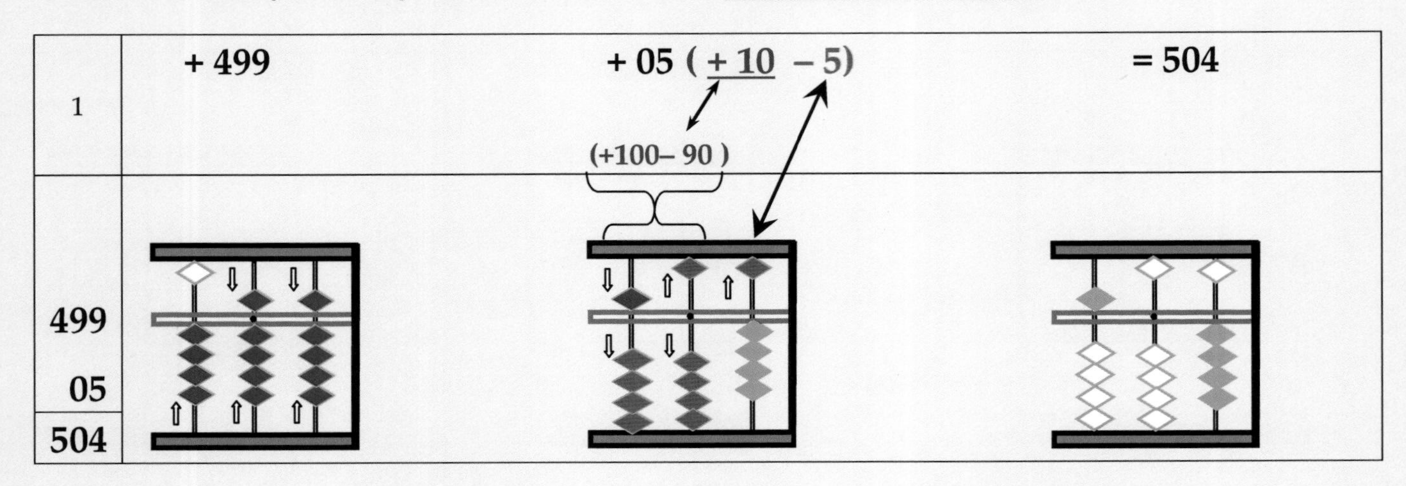

Problem	
+ 499	Move **four earth beads** to touch the beam on the **hundreds rod**. Move the **heaven bead and four earth beads** to touch the beam on the **tens rod.** Move the **heaven bead and four earth beads** to touch the beam on the **ones rod.**
+ 05	There is nothing to do on the tens rod because tens place number is zero. Now we need to +5 on the ones rod, however we do not have enough beads on the ones rod to +5. So, now we need to use the big friend formula of **+5 = +10 – 5** ***Here, all the beads on the tens place rod are in the game, so we cannot do +10 directly.*** ***Now to do +10, you GET HELP from BIG FRIEND FORMULA of +10 and complete the step.*** *Big friend formula for +10 = +100 – 90* However, ***here there are not enough earth beads to +100. Now, GET HELP from small friend formula of* +100 = +500 – 400.** **Now, +5 = +10 (+100 (+500 – 400) – 90) – 5** **Step 1: Add 10** – **Move the heaven bead down to touch the beam and four earth beads down to touch the frame on the hundreds rod to do +100 AND move all the beads away from the beam on the tens rod to complete – 90.** **Step 2: Minus 5** – Move the **heaven bead away from the beam on the ones rod.** ***(Be sure to make students understand that – 5 (big friend of 5 is 5) has to be done on the ones rod because we are trying to +5 on the ones rod.)***

EXAMPLES 4: 495 + 08 = 503 (+8 = +10 – 2)

In this example you will not be able to +8 directly. In such situations get help from appropriate small and big friend formulas to finish the original formula of +8.

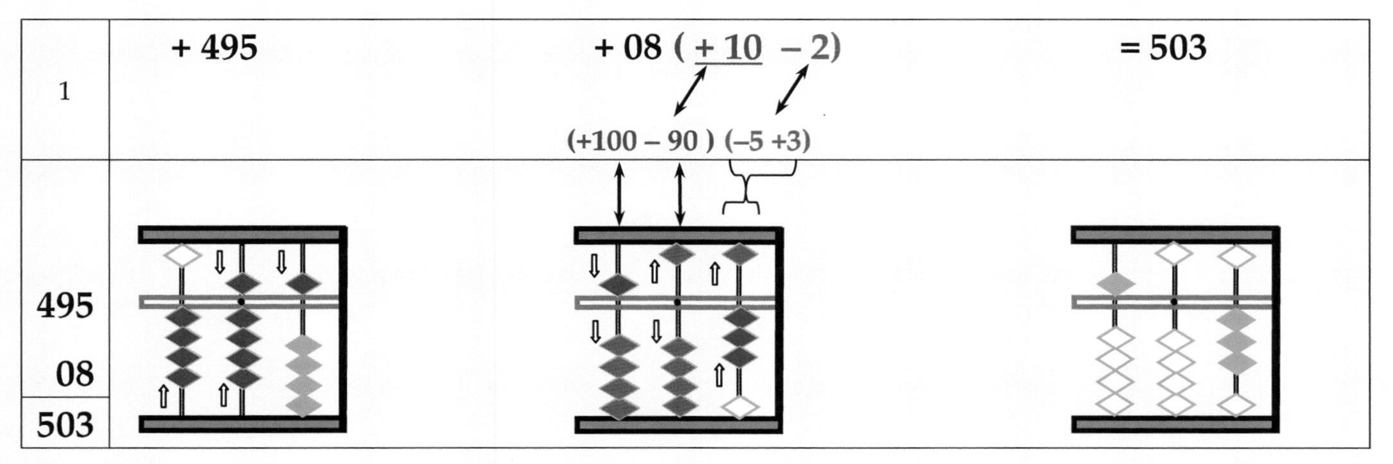

Problem	
+ 495	Move **four earth beads** to touch the beam on the **hundreds rod**. Move the **heaven bead and four earth beads** to touch the beam on the **tens rod.** Move the **heaven bead** to touch the beam on the **ones rod.**
+ 08	There is nothing to do on the tens rod because tens place number is zero. Now we need to +8 on the ones rod, however we do not have enough beads on the ones rod to +8. So, now we need to use the big friend formula of +8 = +10 – 2 ***Here, all the beads on the tens place rod are in the game, so we cannot do +10 directly.*** ***Now to do +10, you GET HELP from BIG FRIEND FORMULA of +10 and complete the step.*** ***Big friend formula for +10 = +100 – 90*** However, ***here there are not enough earth beads to +100. Now, GET HELP from small friend formula of +100 = +500 – 400.*** **Now, +8 = +10 (+100 (+500 – 400) – 90) – 2** Step 1: Add 10 – **Move the heaven bead down to touch the beam and four earth beads down to touch the frame on the hundreds rod to do +100 AND move all the beads away from the beam on the tens rod to complete – 90.** Step 2: Minus 2 – **Move the heaven bead up to touch the frame on the ones rod and move three earth beads up to touch the beam on the ones rod.** ***(Be sure to make students understand that –2 has to be done on the ones rod because we are trying to +8 on the ones rod.)***

ATTENTION

This week's concept may be challenging to younger children. With tenacity and patience, children will master this concept.
The following may help the younger students and those who have hard time understanding the concept:

- Ask them to hold a finger on the frame under the rod (working rod) where they need to add a number and try to solve the problem. This will help them to keep track of where they need to add or subtract.

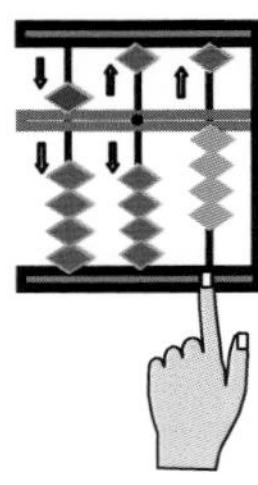

- Ask students to say the formula while they use it. This makes it easy for them to understand and follow through with all the steps. This will also ensure that they use the correct formula.

- In situations where they **get help** from small or big friend formulas, students should not be saying those formulas.

If the above idea does not help, you can ask them to place their finger under the rod they need to add and 'add one' on any higher place value rod wherever possible and clear all rods in between. Then they must finish with the rest of the formula on the rod above their finger. This is a short cut to doing this week's concept. *Please use this method at your own discretion.*

One common mistake almost all students make is double adding.

Children are used to doing +10 on the tens place rod all through this level. However, from this lesson they learn to substitute +10 with +100 – 90 in certain situations.

In such case, due to habit, even though they +100 – 90 to do +10, children will absent mindedly go back and add another ten on the tens rod.

Eg: **99 + 5 = +10** (+100 – 90) **– 5 = 104**

Students will do the above steps as per this week's concept, however due to force of habit they will go back and add another ten on the tens rod and arrive at 114 as their answer.

Ask them to say the formula as they move the required beads to keep themselves on track and avoid doing this mistake. With patience and practice this mistake can be overcome successfully.

LESSON 15 – SAMPLE PROBLEMS

+ 1 = + 10 – 9 **+ 10 = + 100 – 90**

Repeat problems 6 to 10 given under Lesson 14 – Sample Problems in rows 15:S:1 and 15:S:2 before working on the rows below.

Work these problems a few times to study and understand the concept and the relationship between the beads moved.

For Example 1:

1	2	3	4	5	6	7	8	9	10	
99	99	99	99	99	99	99	99	99	49	16:S:1
01	02	03	04	05	06	07	08	09	56	

© SAI Speed Math Academy, USA

For Example 2:

1	2	3	4	5	6	7	8	9	10	
95	95	96	95	96	97	95	96	97	98	16:S:2
09	08	08	07	07	07	06	06	06	06	

© SAI Speed Math Academy, USA

For Example 3:

1	2	3	4	5	6	7	8	9	10	
499	299	199	499	399	499	299	499	199	149	16:S:3
01	202	303	04	105	06	207	08	309	354	

© SAI Speed Math Academy, USA

For Example 4:

1	2	3	4	5	6	7	8	9	10	
495	195	296	395	496	397	295	196	497	298	16:S:4
09	308	208	107	07	107	206	306	06	206	

© SAI Speed Math Academy, USA

POINTS TO REMEMBER

The rows above consist of sample problems to introduce this week's formula. Explain to your child when and how to use the formula. Work with the sample problems until your child understands the formula and that the formulas are to be used ONLY when there are not enough beads to add or subtract.

LESSON 15 – PRACTICE PROBLEMS

We have given a few problems in the rows below where students have had the most difficulty.

1	2	3	4	5	6	7	8	9	10	16:S:5
99 01	99 11	89 01	89 11	99 02	99 12	89 12	89 02	99 03	99 13	

© SAI Speed Math Academy, USA

1	2	3	4	5	6	7	8	9	10	16:S:6
89 03	89 13	99 04	99 14	89 04	89 14	99 05	99 15	89 15	89 05	

© SAI Speed Math Academy, USA

1	2	3	4	5	6	7	8	9	10	16:S:7
99 06	99 16	89 06	89 16	99 07	99 17	89 17	89 07	99 08	99 18	

© SAI Speed Math Academy, USA

1	2	3	4	5	6	7	8	9	10	16:S:8
89 08	89 18	99 09	99 19	89 19	89 09	95 09	95 19	85 09	85 19	

© SAI Speed Math Academy, USA

1	2	3	4	5	6	7	8	9	10	16:S:9
96 08	95 18	86 18	85 08	95 07	96 17	87 07	85 17	96 06	88 16	

© SAI Speed Math Academy, USA

1	2	3	4	5	6	7	8	9	10	16:S:10
386 118	405 98	246 258	435 168	395 207	296 117	187 316	485 117	206 296	488 116	

© SAI Speed Math Academy, USA

Dear Parents and Teachers,

We have come to the end of this level. By now, students should be able to comfortably calculate two digit numbers on abacus and in mind. Most students will show a good understanding of the concepts and will also be able to attain good speed and accuracy.

As parents and teachers, you will be able to better judge the knowledge level of your student. If you feel your student needs more practice before taking the 'End of Level' test, you may choose to make your own worksheets for your students or go to **www.abacus-math.com** and utilize the free worksheet generator to build and print worksheets for your students.

When you generate worksheet from the above website, please choose the values for the criterias as given in the picture.

Free Worksheet builder x

MATH PROBLEMS

Number of values per question:	6
Number of Sums:	12
Minimum number of digits:	2
Maximum number of digits:	2
Select Sheet:	Addition

SUBMIT

Once students are fluent working with the worksheets, then you may choose to increase the values for the criterias and print more challenging work.

You can let them take the End of Level test given in the LEVEL 2 – WORKBOOK 2 when your student is ready.

Good Luck with the test!

Thank you,

SAI Speed Math Academy

EXCLUSIVE BONUS INTERMEDIATE LEVEL LESSONS

NOTE TO PARENTS AND TEACHERS

Dear Parents and Teachers,

We hope that by now your students are working with ease on abacus and in mind and have successfully completed their End of Level Test found at the end of the LEVEL 2 – WORKBOOK – 2.

The following two lessons teach about using the concepts learned in this level on the hundreds rod. This is a special bonus section given ONLY in this LEVEL 2 – INSTRUCTION BOOK. These two lessons are at intermediate level in intensity. Young children may find them slightly challenging. Please work on these two lessons only when your student masters working with two digit numbers. As parents and teachers, you will be able to better judge the knowledge level of your student. You can decide according to your child's readiness to introduce these lessons at the appropriate time.

After finishing the work given here you may make your own practice worksheets or you may choose to go to **www.abacus-math.com** and utilize the free worksheet generator to build and print worksheets for your students.

Thank you,
SAI Speed Math Academy

WEEK 21 – LESSON 16 – INTRODUCING – USING BIG FRIENDS FORMULAS ON HUNDREDS PLACE ROD

CONCEPTS OF THE WEEK

TO ADD A HUNDREDS PLACE NUMBER = ADD 1000, LESS BIG FRIEND

ADDING ON HUNDREDS PLACE ROD

By now students would have understood the theory behind the big friend formulas. Students would have successfully learned to use the formulas on the ones place rod and tens place rod. Students will easily adapt those formulas to successfully add on the hundreds place rod.

Concept for using the formula on the hundreds place rod is: If you do not have enough beads to add on the hundreds place rod, **'add 1000' on the thousands place rod** and **send out the big friend** of the number being added from the beam **on the hundreds place rod**.

When teaching younger children, you can ask them to hold their finger under the hundreds rod where they are adding and 'add' on the higher place value rod 'on the left' (thousands place rod) and 'minus the friend' of the number being added from the working rod.

The practice work corresponding to this lesson and the next lesson is not given in the workbook. You can choose to teach these two concepts when you think your student is ready to take on more challenging work, and then allow them to do the practice work for abacus and mind math that is given here after the sample work. Some students will take more time to get to this stage than others. Please make sure that your students are not challenged too much, too soon.

BIG FRIEND COMBINATION FACTS FOR 1000

100 and 900 are big friends of 1000, because together they make 1000.
200 and 800 are big friends of 1000, because together they make 1000.
300 and 700 are big friends of 1000, because together they make 1000.
400 and 600 are big friends of 1000, because together they make 1000.
500 and 500 are big friends of 1000, because together they make 1000.

+900 = +1000 – 100	**+400 = +1000 – 600**
+800 = +1000 – 200	**+300 = +1000 – 700**
+700 = +1000 – 300	**+200 = +1000 – 800**
+600 = +1000 – 400	**+100 = +1000 – 900**
+500 = +1000 – 500	**+1000 = +10000 – 9000**

LESSON 16 – EXAMPLE

EXAMPLE 1: 499 + 900 = 1399 (+900= +1000 – 100)

1	+ 499	+ 900 (+ 1000 – 100)	= 1399
499 900 1399	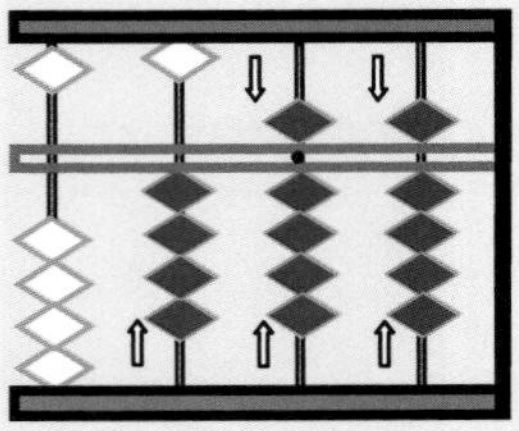	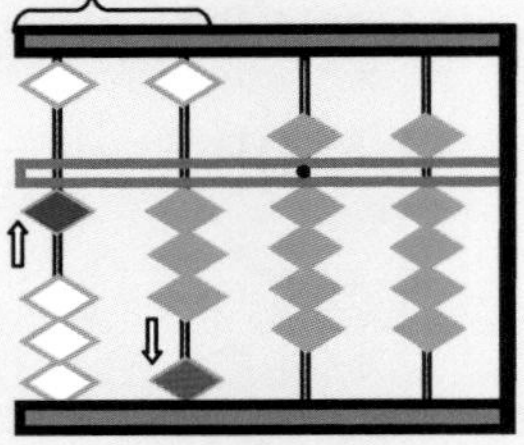	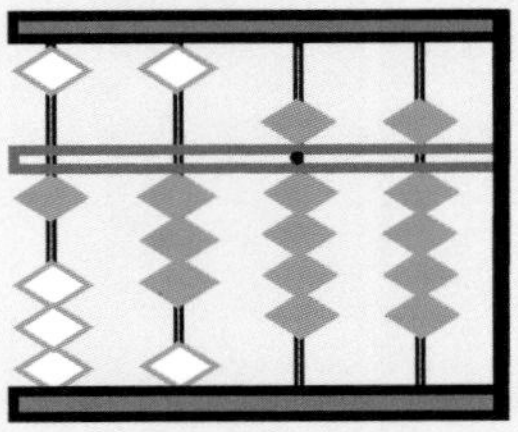

Problem	Action
+ 499	Set the numbers on their appropriate place value rods.
+ 900	Now we need to **+900** on the hundreds rod, however we do not have enough beads on the hundreds rod to +900. So, now we need to make use of the fact that **900 + 100 = 1000** **When you want to +900 and you do not have enough beads:** **Use +900 = +1000 – 100 Big Friend formula of hundreds numbers to do your calculations.** **Step 1: Add 1000** – Move **one earth bead up** to touch the beam on the **thousands rod**. *(One earth bead on the thousands rod is equal to '1000')* *We know that there is a nine hundred in the thousand (900 + 100 = 1000), so let us get help from 1000 by adding it to our game. However, we were supposed to +900, instead we did +1000, which means we have 100 more than what we need. So, now we have to send the 100 away from our game.* **Step 2: Minus 100** – Move **one earth bead away** from the beam on the **hundreds rod**. ***(When we do + 1000 and – 100, we get to keep 900 in our game.)*** There is nothing to do on the tens place rod because tens place number is zero. There is nothing to do on the ones place rod because ones place number is zero

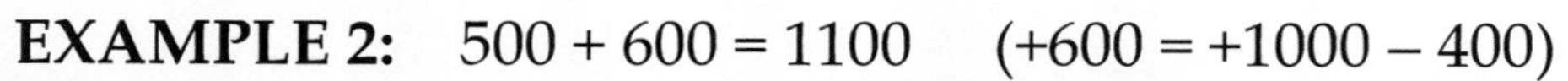

EXAMPLE 2: 500 + 600 = 1100 (+600 = +1000 – 400)

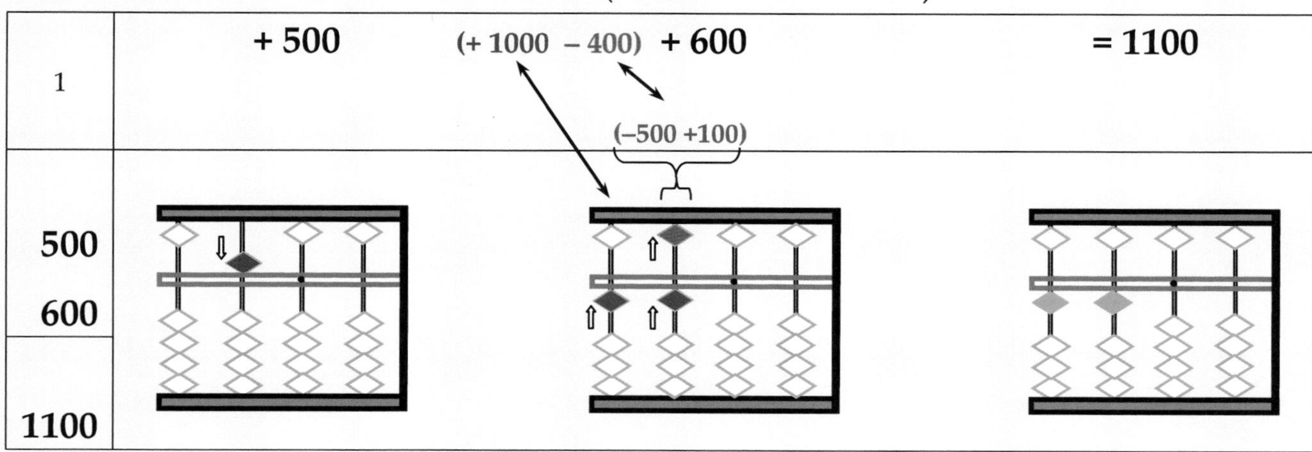

Problem	Action
+ 500	Set the number on the appropriate place value rod.
+ 600	Now we need to +600 on the hundreds rod, however we do not have enough beads on the hundreds rod to +600. So, now we need to make use of the fact that **600 + 400 = 1000** **When you want to +600 and you do not have enough beads:** **Use +600 = +1000 – 400 Big Friend formula of hundreds numbers to do your calculations.** **Step 1: Add 1000** – Move one earth bead up to touch the beam on the **thousands rod.** *(One earth bead on the thousands rod is equal to '1000')* ***Now we need to finish the formula by doing – 400 on the hundreds rod, but we do not have earth beads to do – 400 so, GET HELP from the small friend formula of – 4 and use it on the thousands rod to do – 400.*** **Step 2: Minus 400** = **Step A: Minus 500** – Move the heaven bead up to touch the frame on the hundreds rod. **Step B: Add 100** – Move one earth bead up to touch the beam on the hundreds rod. ***(When we do + 10 and – 4, we get to keep 6 in our game.)*** There is nothing to do on the tens place rod because tens place number is zero. There is nothing to do on the ones place rod because ones place number is zero

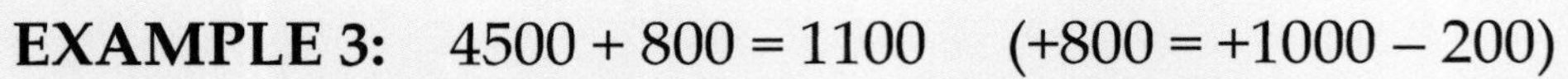

EXAMPLE 3: 4500 + 800 = 1100 (+800 = +1000 – 200)

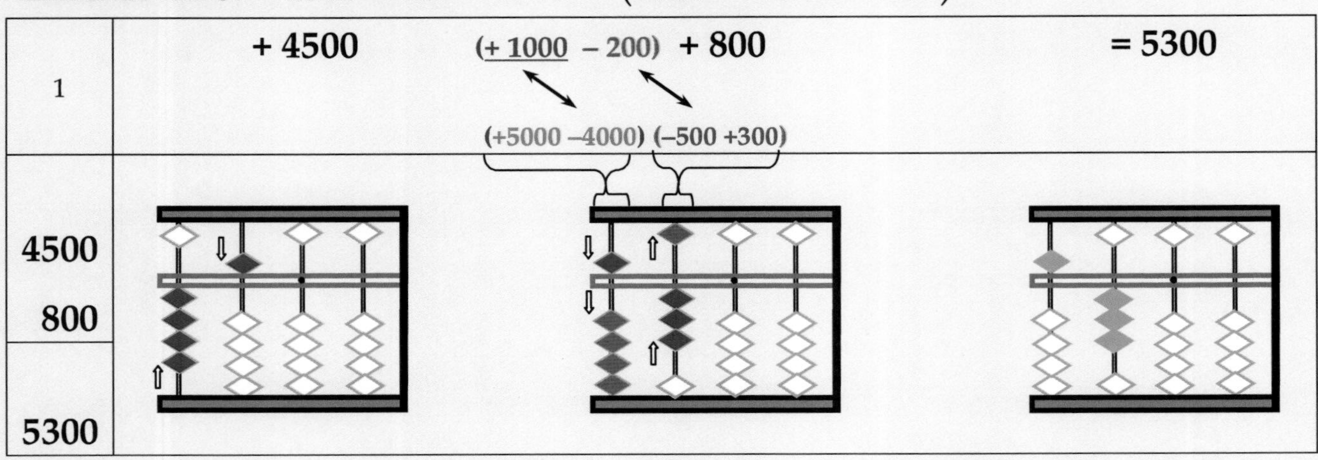

Problem	Action
+ 4500	Set the number on the appropriate place value rods.
+ 800	Now we need to +800 on the hundreds rod, but we do not have enough beads on the hundreds rod to +800. So, now we need to make use of the fact that **800 + 200 = 1000** **When you want to +800 and you do not have enough beads:** **Use +800 = +1000 – 200 Big Friend formula of hundreds numbers to do your calculations.** ***Now to do +1000 you do not have enough beads so, GET HELP from small friend formula of +1 and use it on the thousands rod to do +1000.*** **Step 1: Add 1000 =** **Step A: Add 5000** – Move the heaven bead down to touch the beam on the thousands rod. **Step B: Minus 4000** – Move all four earth beads down to touch the frame on the thousands rod. ***Now we need to finish the formula by doing – 200 on the hundreds rod, but we do not have enough earth beads to do –200 so, GET HELP from small friend formula of – 2 and use it on the hundreds rod to do – 200.*** **Step 2: Minus 200 =** **Step A: Minus 500** – Move the heaven bead up to touch the frame on the hundreds rod. **Step B: Add 300** – Move three earth beads up to touch the beam on the hundreds rod. ***(When we do + 1000 and – 200, we get to keep 800 in our game.)*** There is nothing to do on the tens place rod because tens place number is zero. There is nothing to do on the ones place rod because ones place number is zero

ATTENTION

This week's concept may be challenging for younger children. With tenacity and patience, children will master this concept. Please make sure they are not distracted while working.

The following may help the younger students and those who have hard time understanding the concept:

- Ask them to hold a finger on the frame under the rod (working rod) where they need to add a number and try to solve the problem. This will help them to keep track of where they need to add or subtract.

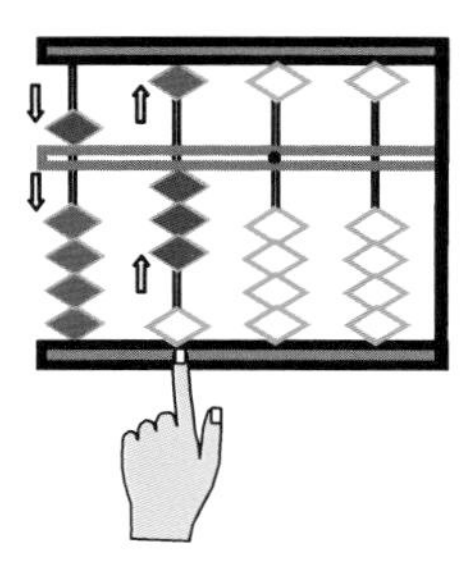

- Ask students to say the formula while they use it. This makes it easy for them to understand and follow through with all the steps. This will also ensure that they use the correct formula.
- In situations where they **get help** from small or big friend formulas, students should not be saying those formulas.

Keeping track of the place value of rods and adding or subtracting numbers on their corresponding place value rod is the key to success.

LESSON 16 – SAMPLE PROBLEMS

Work these problems a few times to study and understand the concept and the relationship between the beads moved.

For Example 1:

1	2	3	4	5	6	7	8	9	10
700	805	901	823	315	950	803	641	539	655
900	800	100	202	705	606	333	414	511	987

21:S:1

© SAI Speed Math Academy, USA

For Example 2:

1	2	3	4	5	6	7	8	9	10
500	606	725	805	761	883	999	551	608	508
900	805	739	646	674	243	103	789	747	811

21:S:2

© SAI Speed Math Academy, USA

For Example 3:

1	2	3	4	5	6	7	8	9	10
4900	4505	4605	4999	2500	3804	1908	1647	4199	3599
100	808	795	203	2600	1444	3396	3555	904	855

21:S:3

© SAI Speed Math Academy, USA

Sample Problems

1	2	3	4	5	6	7	8
385	900	769	334	594	977	128	166
364	195	156	376	839	131	774	471
898	407	579	480	104	775	544	665

21:S:4

© SAI Speed Math Academy, USA

1	2	3	4	5	6	7	8
934	836	759	654	569	804	591	149
420	290	651	194	397	770	764	355
733	522	755	365	735	225	395	536
- 62	- 408	- 34	396	- 501	06	- 410	494

21:S:5

© SAI Speed Math Academy, USA

LESSON 16 – PRACTICE WORK

Use Abacus

DAY 1 – MONDAY

TIME: ___________min ______________sec Accuracy _______/16

1	2	3	4	5	6	7	8
644	588	424	743	940	591	105	662
97	529	277	565	941	516	347	427
233	266	555	- 202	- 521	649	- 440	606
860	878	215	396	553	- 444	488	- 555

21:1

© SAI Speed Math Academy, USA

1	2	3	4	5	6	7	8
3486	1162	2696	1695	8486	2948	7389	4442
138	407	887	207	161	656	855	145
276	286	998	709	743	1446	258	914

21:2

© SAI Speed Math Academy, USA

DAY 2 – TUESDAY

TIME: ___________min ______________sec Accuracy _______/16

1	2	3	4	5	6	7	8
215	438	856	585	534	986	181	285
554	652	79	36	42	820	968	696
- 235	563	369	514	- 230	- 505	257	- 160
- 130	742	268	738	155	678	896	666

21:3

© SAI Speed Math Academy, USA

1	2	3	4	5	6	7	8
2158	2377	2376	2366	4838	1780	8759	6786
615	138	377	189	354	855	- 412	- 5010
2511	5881	1549	2945	115	1709	159	3726

21:4

© SAI Speed Math Academy, USA

DAY 3 – WEDNESDAY

TIME: ________min ________sec Accuracy ______/16 ☆

1	2	3	4	5	6	7	8
918	398	492	920	123	937	476	752
357	216	59	179	31	48	115	763
605	352	932	404	599	- 601	- 30	- 401
401	555	233	224	462	528	884	66

21:5

© SAI Speed Math Academy, USA

1	2	3	4	5	6	7	8
2959	3535	4219	4655	3248	3714	7456	6490
600	446	656	868	568	476	- 3040	766
1592	1188	325	503	1674	319	688	1555

21:6

© SAI Speed Math Academy, USA

DAY 4 – THURSDAY

TIME: ________min ________sec Accuracy ______/16 ☆

1	2	3	4	5	6	7	8
327	772	619	852	253	817	352	977
557	315	943	896	377	349	331	13
827	113	705	- 621	219	- 124	65	129
689	954	535	475	202	418	524	388

21:7

© SAI Speed Math Academy, USA

1	2	3	4	5	6	7	8
			3329	9745	3729	2724	1691
5280	4224	3333	637	- 6301	172	334	414
156	857	167	1151	1151	1104	428	894
2383	2166	1901	387	520	- 5005	718	152

21:8

© SAI Speed Math Academy, USA

DAY 5 – FRIDAY

TIME: __________min ____________sec Accuracy _______/16

1	2	3	4	5	6	7	8
632	159	922	926	426	227	705	987
821	356	323	111	527	279	614	247
298	244	692	518	866	706	193	267
647	- 303	- 525	796	906	398	983	855
110	111	353	287	705	505	14	198

21:9

© SAI Speed Math Academy, USA

1	2	3	4	5	6	7	8
			1165	5654	3686	966	4082
3959	7943	7887	212	754	293	4756	358
626	453	- 5401	173	- 4404	1122	- 2010	549
1919	1603	753	3959	196	816	1538	213

21:10

© SAI Speed Math Academy, USA

SEPARATE SHAPES

Draw 3 lines to separate 3 diamonds into their own group. Each group must have 3 diamonds in the space.

Example:

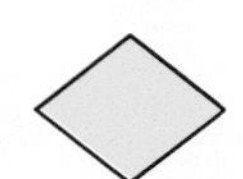
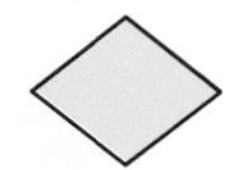
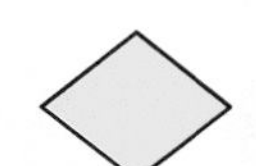
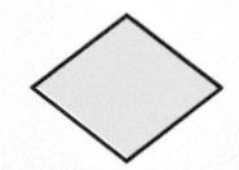
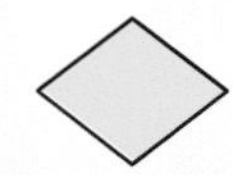
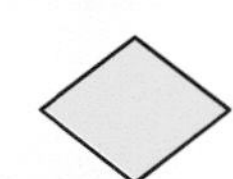
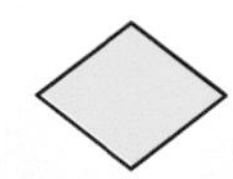
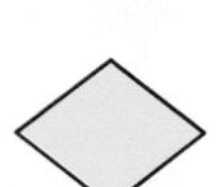
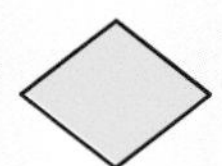
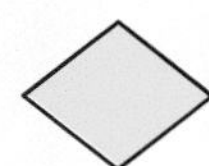

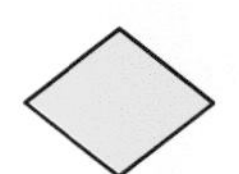

LESSON 16 – MIND MATH PRACTICE WORK

Accuracy ______/10

DAY 1 – MONDAY

1	2	3	4	5	6	7	8	9	10
1056	3963	2321	3184	4555	2273	496	1160	677	2387
541	110	580	370	511	707	605	195	850	522

21:11

© SAI Speed Math Academy, USA

Accuracy ______/10

DAY 2 – TUESDAY

1	2	3	4	5	6	7	8	9	10
198	772	2918	5641	858	2566	678	645	494	587
902	508	102	-1401	603	134	850	734	607	934

21:12

© SAI Speed Math Academy, USA

Accuracy ______/10

DAY 3 – WEDNESDAY

1	2	3	4	5	6	7	8	9	10
1724	542	6737	985	3966	732	963	914	3756	1469
230	619	-5206	425	303	590	246	196	45	31

21:13

© SAI Speed Math Academy, USA

Accuracy ______/10

DAY 4 – THURSDAY

1	2	3	4	5	6	7	8	9	10
					4600	344	500	949	360
3544	1515	1554	4505	306	705	554	808	120	487
1600	505	48	809	706	-3303	502	44	35	307

21:14

© SAI Speed Math Academy, USA

DAY 5 – FRIDAY Accuracy _______/20 ☆

1	2	3	4	5	6	7	8	9	10
536 504	5115 503	1554 48	555 945	3006 708	996 105	2026 733	833 221	799 111	1909 341

21:15

© SAI Speed Math Academy, USA

1	2	3	4	5	6	7	8	9	10
745 09 47	46 466 722	229 601 808	71 514 820	940 270 3800	569 82 550	265 393 44	4004 550 901	1704 700 3151	474 180 528

21:16

© SAI Speed Math Academy, USA

LESSON 16 – DICTATION

DICTATION: Dictation is when teacher or parent calls out a series of numbers and the child listens to the numbers and does the calculation in mind or on the abacus.

DO 6 PROBLEMS A DAY and write answers below.
Students: Try to calculate the dictated problems in mind.
Teachers: Dictate problems from mind math part of this week's work.

1	2	3	4	5	6	7	8	9	10

11	12	13	14	15	16	17	18	19	20

21	22	23	24	25	26	27	28	29	30

WEEK 22 – LESSON 17 – INTRODUCING – REGROUPING OR CARRYOVER TO THOUSANDS ROD

BIG FRIEND COMBINATION FACTS FOR 1000

100 and 900 are big friends of 1000, because together they make 1000.

200 and 800 are big friends of 1000, because together they make 1000.

300 and 700 are big friends of 1000, because together they make 1000.

400 and 600 are big friends of 1000, because together they make 1000.

500 and 500 are big friends of 1000, because together they make 1000.

+900 = +1000 – 100	**+400 = +1000 – 600**
+800 = +1000 – 200	**+300 = +1000 – 700**
+700 = +1000 – 300	**+200 = +1000 – 800**
+600 = +1000 – 400	**+100 = +1000 – 900**
+500 = +1000 – 500	**+1000 = +10000 – 9000**

LESSON 17 – EXAMPLE

EXAMPLE 1: 990 + 10 = 1000 **(+10= +100** (+1000 – 900) **– 90)**

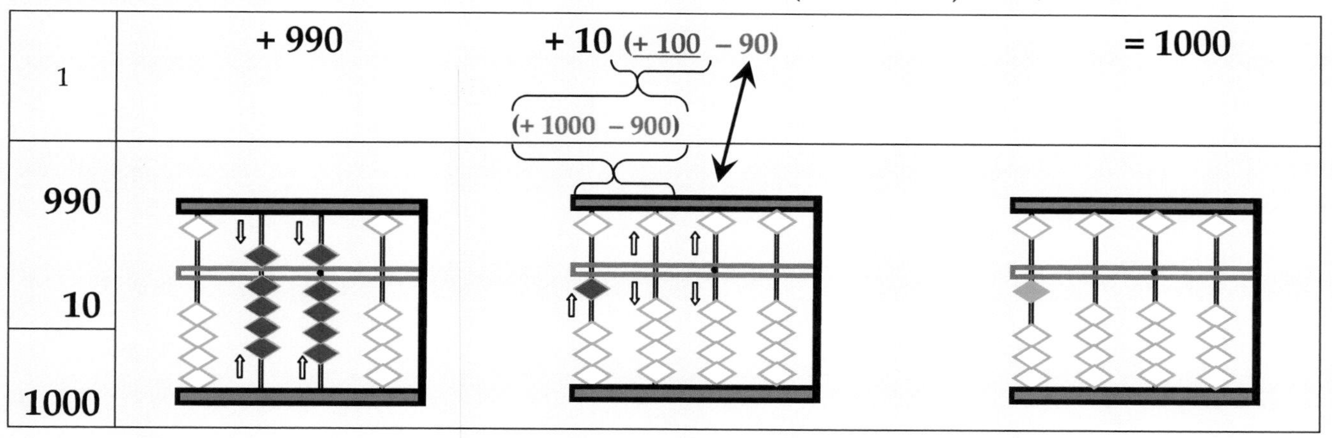

Problem	Action
+ 990	Set the numbers on their appropriate place value rods.
+ 10	Now we need to **+10** on the tens rod, however we do not have enough beads on the tens rod to +10. So, now we need to make use of the fact that **10 + 90 = 100** **When you want to +10 and you do not have enough beads:** **Use +10 = +100 – 90 Big Friend formula of hundreds numbers to do your calculations.** **Step 1: Add 100 =** ***Here, all the beads on the hundreds place rod are already in the game, so we cannot do +100 directly. Now to do +100, you GET HELP from BIG FRIEND FORMULA of +100 and complete the step.*** ***Big friend formula for +100 = +1000 – 900*** **Now, +10 = +100** ***(+1000 – 900)*** **– 90** **Step 1: Add 100** **Step A: Add 1000** **Move one earth bead up on the thousands rod** *(One earth bead on the thousands rod is equal to '1000')* **Step B: Minus 900** **by moving the heaven bead and all earth beads away from the beam on the hundreds rod.** **Step 2: Minus 90** – Move **the heaven bead and all earth beads away from the beam on the tens rod.** There is nothing to do on the ones place rod because ones place number is zero

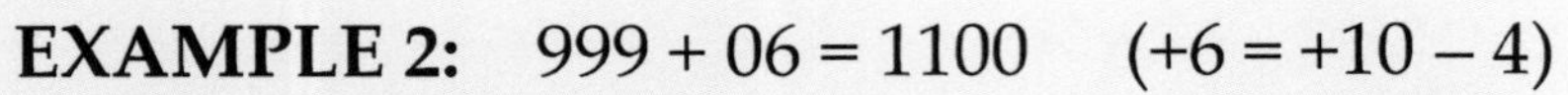

EXAMPLE 2: 999 + 06 = 1100 (+6 = +10 – 4)

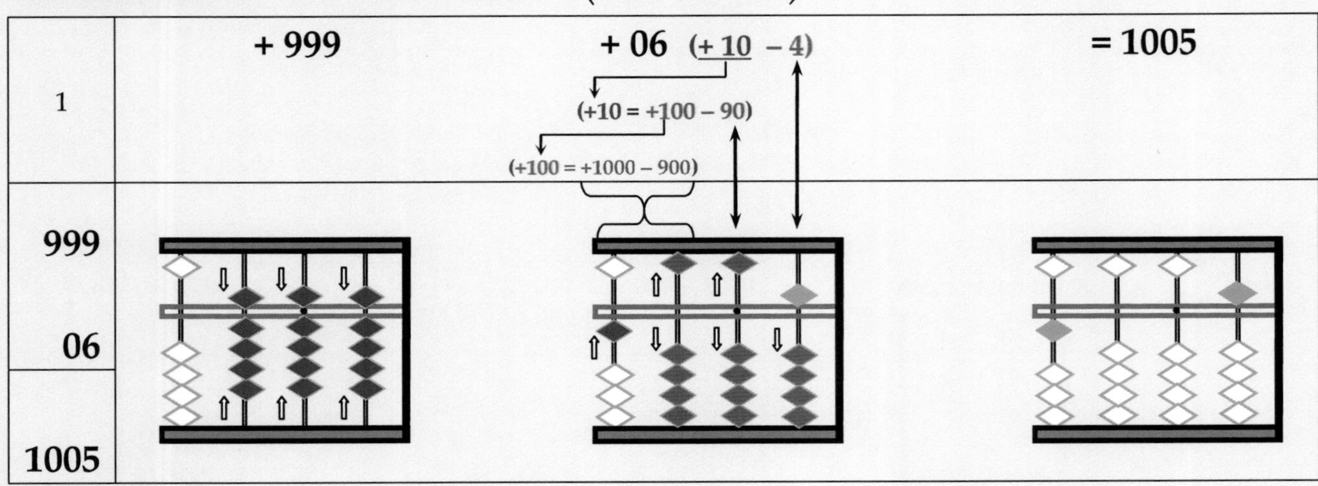

Problem	Action
+ 999	Set the number on the appropriate place value rod.
+ 06	Now we need to +6 on the ones rod, however we do not have enough beads on the ones rod to +6. So, now we need to make use of the fact that **6 + 4 = 10** **When you want to +6 and you do not have enough beads:** **Use +6 = +10 – 4 Big Friend formula of hundreds numbers to do your calculations.** **Step 1: Add 10 =** ***Here, all the beads on the tens place rod are already in the game, so we cannot do +10 directly. Now to do +10, you GET HELP from BIG FRIEND FORMULA of +10 and complete the step. Big friend formula for +10 = +100 – 90 However, here the hundreds place rod is also full. Now to do +100, GET HELP from BIG FRIEND FORMULA of +100 = +1000 – 900*** **Now, +6 = +10** ***(+100(+1000 – 900) – 90)*** **– 4** **Step 1: Add 10** **Step A: Add 100** **Step a: Add 1000 –** Move **one earth bead up** to touch the beam on the **thousands rod**. *(One earth bead on the thousands rod is equal to '1000')* **Step b: Minus 900** – Move the **heaven bead and all the earth beads away** from the beam on the **hundreds rod**. **Step A: Minus 90** – Move the **heaven bead and all the earth beads away** from the beam on the **tens rod**. ***Now we need to finish the formula by doing – 4 on the ones rod.*** **Step 2: Minus 4** = Move **four earth beads away** from the beam on the **ones rod.**

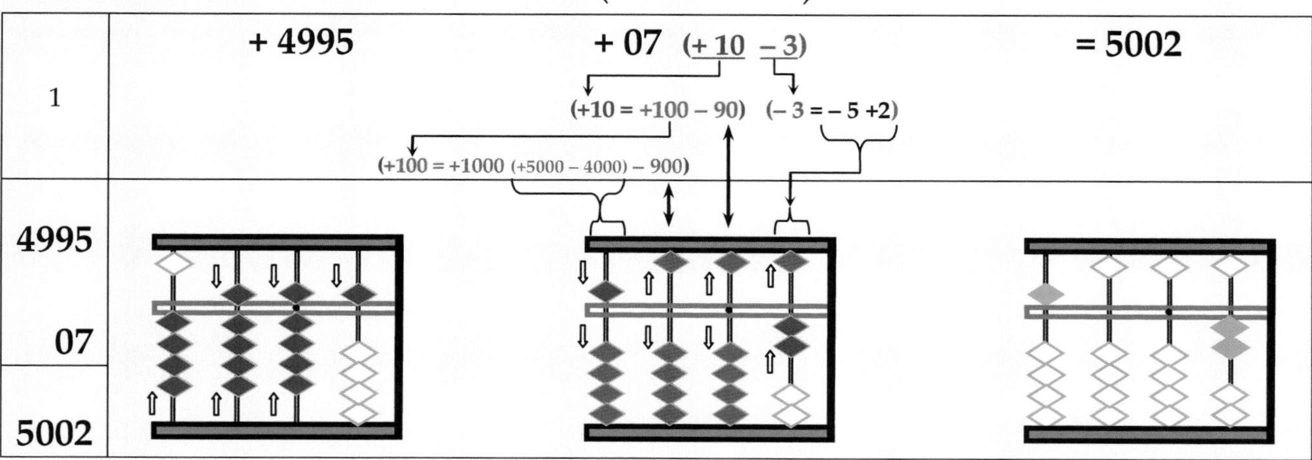

Problem	Action
+ 4995	Set the number on the appropriate place value rod.

+ 07	Now we need to +7 on the ones rod, however we do not have enough beads on the ones rod to +7. So, now we need to make use of the fact that **7 + 3 = 10**
	When you want to +7 and you do not have enough beads:
	Use +7 = +10 – 3 Big Friend formula of hundreds numbers to do your calculations.
	Step 1: Add 10 = *Here, all the beads on the tens place rod are already in the game, so we cannot do +10 directly. Now to do +10, you GET HELP from BIG FRIEND FORMULA of +10 and complete the step. Big friend formula for* ***+10 = +100 – 90*** *However, here the hundreds place rod is also full. Now to do +100, GET HELP from BIG FRIEND FORMULA of +100 = +1000 – 900. In the above example, you will have to GET HELP from +1000 = +5000 – 4000 SMALL FRIENDS FORMULA to fulfill the requirement of +10 in the +7 = +10 – 3 formula.*
	Now, +7 = +10 *(+100(+1000 (+5000 – 4000) – 900) – 90)* **– 3**
	
	<u>Step 1: Add 10</u>
	    <u>Step A: Add 100</u>
	        <u>Step a: Add 1000</u> = (+5000 – 4000) – Move the **heaven bead down** to touch the beam and move all **4 earth beads down** to touch the frame on the **thousands rod.**
	        <u>Step b: Minus 900</u> – Move the **heaven bead and all the earth beads away** from the beam on the **hundreds rod**.
	    <u>Step A: Minus 90</u> – Move the **heaven bead and all the earth beads away** from the beam on the **tens rod**.
	Now we need to finish the formula by doing – 3 on the ones rod.
	<u>Step 2: Minus 3</u> = Move **heaven bead up** to touch the frame and **two earth bead up** to touch the beam on the **ones rod.**

ATTENTION

This week's concept may look complicated and challenging to anyone in the beginning. With tenacity and patience, everyone can master this concept.

The following may help all those who have hard time understanding the concept:

- Ask them to hold a finger on the frame under the rod (working rod*) where they need to add a number and try to solve the problem. This will help them to keep track of where they need to add or subtract.

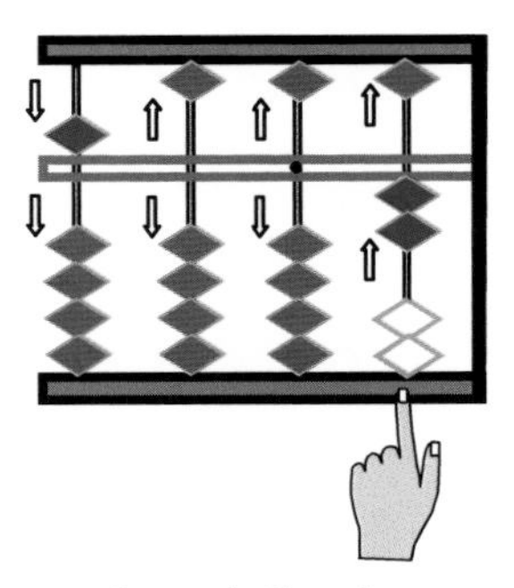

- Ask students to say the formula while they use it. This makes it easy for them to understand and follow through with all the steps. This will also ensure that they use the correct formula.

- In situations where they **get help** from small or big friend formulas, students should not be saying those formulas.

Short Cut Method**

The easiest way to add under the circumstances as shown in the examples is:

<u>Step 1:</u> **Add 'one'** wherever possible, **on** any **higher place value rod** from the working rod*
<u>Step 2:</u> **Minus** the '**big friend** of the number' being added **on** the **'working rod'**

(*working rod is the rod corresponding to the place value of the number being added)
(**Please use this method at your own discretion.)

LESSON 17 – SAMPLE PROBLEMS

Work these problems a few times to study and understand the concept and the relationship between the beads moved.

For Example 1:

1	2	3	4	5	6	7	8	9	10
326	755	951	863	355	954	899	699	384	193
177	280	54	162	695	66	132	311	641	857

22:S:1

© SAI Speed Math Academy, USA

For Example 2:

1	2	3	4	5	6	7	8	9	10
556	686	228	809	768	888	999	988	608	909
444	315	779	191	233	112	06	17	398	95

22:S:2

© SAI Speed Math Academy, USA

For Example 3:

1	2	3	4	5	6	7	8	9	10
1555	4507	3719	4999	2569	3804	1928	4409	4119	3929
445	498	1282	07	2436	1199	3074	591	883	78

22:S:3

© SAI Speed Math Academy, USA

1	2	3	4	5	6	7	8
1785	3806	2065	3305	2595	4977	4988	3166
3219	1198	2937	1696	2405	26	12	848

22:S:4

© SAI Speed Math Academy, USA

Sample Problems

1	2	3	4	5	6	7	8
2323	3414	4758	5668	4075	7362	1292	5829
633	529	246	333	425	145	435	114
253	15	3249	55	28	48	3733	25
996	1045	- 4010	975	474	549	550	36

22:S:5

© SAI Speed Math Academy, USA

LESSON 17 – PRACTICE WORK

Use Abacus

DAY 1 – MONDAY

TIME: __________min ____________sec Accuracy ______/16

1	2	3	4	5	6	7	8
934	136	759	654	569	804	595	188
- 420	290	- 657	194	391	74	- 164	288
- 101	522	144	- 315	- 730	- 843	395	36
33	- 704	58	- 311	19	232	822	- 400
55	51	51	29	52	132	51	55
- 301	207	645	780	91	101	301	387

22:1

© SAI Speed Math Academy, USA

1	2	3	4	5	6	7	8
5028	4593	2690	4905	6463	9778	5900	2824
443	846	814	99	580	- 6643	306	146
545	2546	1184	1285	962	360	94	29
1987	18	3782	- 4057	999	1505	- 6300	2001

22:2

© SAI Speed Math Academy, USA

WORD SEARCH

```
U R E E H M A H E R D N A A L A L P
G R A S M A A S E S S H E E P P A F
T A O C R A I P S K H O R E D G N A
C T P I A L P I E N E G I L R A D L
O L A F F U B R E I E K K L I L A L
W I A S A R O O R L A H C O E R A M
R P H O B I R E T S O O R I N S A S
O R G G D M S T O G O H F G H E S S
O E I L D D A U O O C O I R D C T A
S T P A N C J L T R E S P E I G P R
B U U F F L O G H A E S R O H T D G
```

CHICKEN
HEN
ROOSTER
LAMB
SHEEP
HORSE
COW
BUFFALO
CATTLE
HERD
PIG
HOG
FARM
LAND
GRASS
TREES
FIELD

DAY 2 – TUESDAY

TIME: __________min __________sec Accuracy ______/16 ☆

1	2	3	4	5	6	7	8
590	457	744	143	697	423	383	118
47	223	38	785	- 433	32	- 201	584
112	175	- 562	- 307	81	168	405	603
151	97	55	- 401	12	- 321	554	52
- 400	107	14	36	44	92	22	149
- 100	951	14	295	- 401	11	55	497

22:3

© SAI Speed Math Academy, USA

1	2	3	4	5	6	7	8
1482	2423	1997	2565	985	1240	51	4985
1200	467	3003	512	212	358	764	168
1704	2585	99	1728	308	407	3052	874
1762	108	404	298	3499	2997	133	475

22:4

© SAI Speed Math Academy, USA

MIRROR IMAGE

Draw mirror image of the picture on the other side of the line next to it.

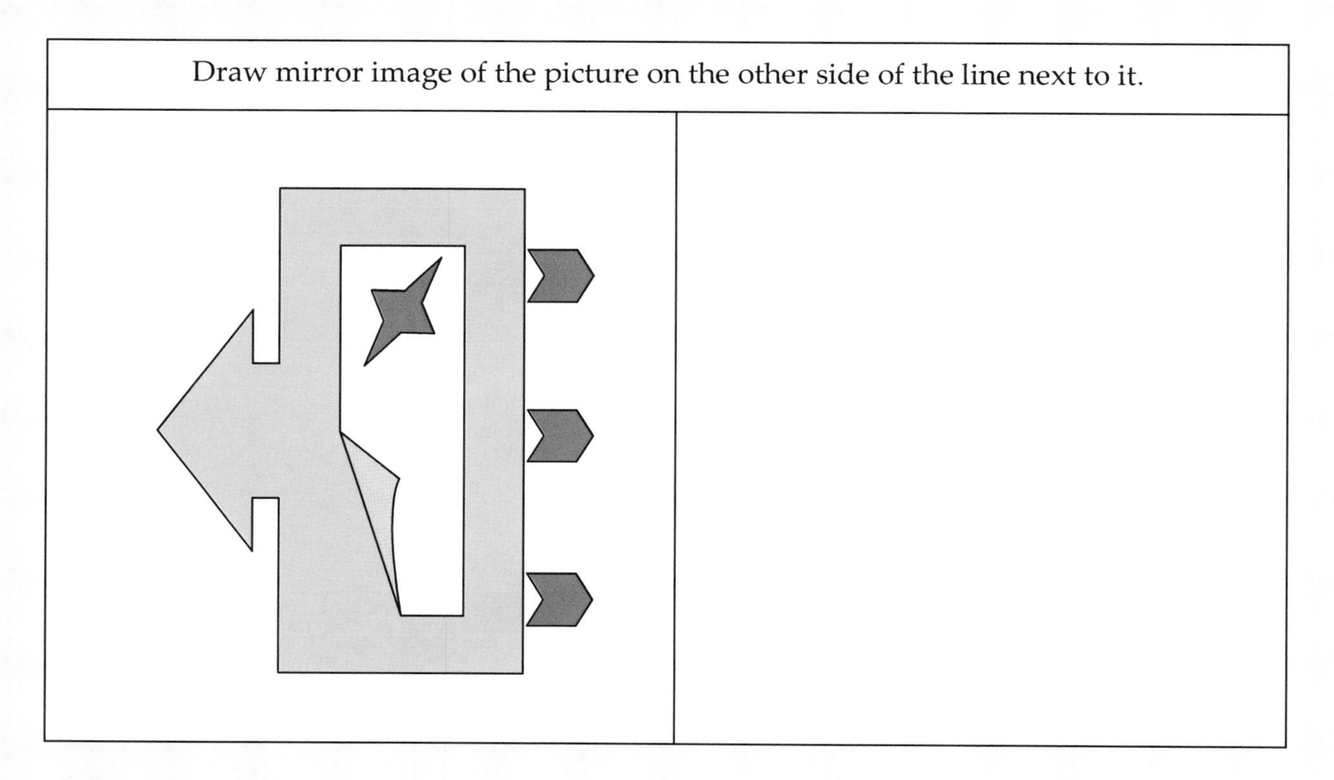

DAY 3 – WEDNESDAY

TIME: __________min ____________sec Accuracy _______/16

22:5

1	2	3	4	5	6	7	8
631	442	166	186	305	52	231	152
194	418	133	318	382	587	633	693
99	629	232	41	- 431	181	- 724	35
124	516	51	- 130	272	705	38	- 770
66	44	- 40	82	71	- 301	- 42	22
481	42	256	103	- 598	77	66	42

© SAI Speed Math Academy, USA

22:6

1	2	3	4	5	6	7	8
4949	2136	1379	3442	3075	2195	3197	1580
51	573	1412	373	437	458	24	369
138	491	254	189	- 1101	1876	535	53
364	- 3200	459	- 3003	509	921	1796	2998

© SAI Speed Math Academy, USA

SEPARATE SHAPES

Draw 3 lines to separate 2 boxes per group.

DAY 4 – THURSDAY

TIME: __________min __________sec Accuracy ______/16 ☆

1	2	3	4	5	6	7	8
152	186	39	255	111	868	201	84
195	457	375	158	320	- 542	640	144
172	- 223	213	43	149	79	85	232
94	75	- 602	- 431	286	41	28	- 410
48	18	- 12	56	37	51	123	- 10
- 661	- 111	358	- 81	97	115	425	60

22:7

© SAI Speed Math Academy, USA

1	2	3	4	5	6	7	8
5724	1137	640	2826	4077	5055	7825	4510
271	864	405	333	715	- 3021	669	528
2193	1566	875	840	105	464	- 6262	921
315	1987	77	1050	202	11	1617	43
- 6101	- 1234	3007	451	909	1493	257	- 2002

22:8

© SAI Speed Math Academy, USA

NUMBER BLOCK

				20
	6		1	20
	6	9		27
6			0	18
	8	1	9	23
24	25	25	14	27

Fill in the missing numbers

Use numbers 0 to 9 and fill in the empty cells.

Clues:

1. Rows add up to the total on the right.
2. Columns add up to the total at the bottom.
3. Diagonal cells up to the total shown in the top and bottom cells.
4. Numbers can repeat within a row or column.

				22
	6	8		23
7	5			16
		3	8	25
9		7	4	28
27	24	19	22	14

DAY 5 – FRIDAY

TIME: ________min ___________sec Accuracy ______/16

1	2	3	4	5	6	7	8
58	134	263	310	250	307	228	371
41	392	228	197	255	262	254	31
151	13	144	- 104	- 201	- 147	17	160
19	- 407	- 601	41	65	73	553	94
15	26	22	55	132	06	317	634
221	44	- 15	55	- 401	549	534	716

22:9

© SAI Speed Math Academy, USA

1	2	3	4	5	6	7	8
4086	5890	2998	1876	424	1726	8547	1561
165	732	201	437	71	270	- 4221	485
26	377	6460	3254	09	3010	28	513
224	51	- 9313	987	1153	660	44	548
609	- 6010	4658	460	843	- 4346	02	44

22:10

© SAI Speed Math Academy, USA

SHAPE PUZZLE

Find out what each shape is worth.

Suggestion: Take as many numbers of beans, blocks or coins as the right side of the equation shows and share them on the shapes to the left side of the equation. Hint: here use third equation as a clue.

+ − = 02

+ + + = 34

+ + = 26

+ − =

= ___ = ___ = ___ = ___

WEEK 22 – MIND MATH PRACTICE WORK

Visualize

Accuracy ______/10 ☆

DAY 1 – MONDAY

1	2	3	4	5	6	7	8	9	10
486 15	550 450	670 370	509 491	3050 990	4995 06	4096 910	3666 1334	5450 550	3299 707

22:11

© SAI Speed Math Academy, USA

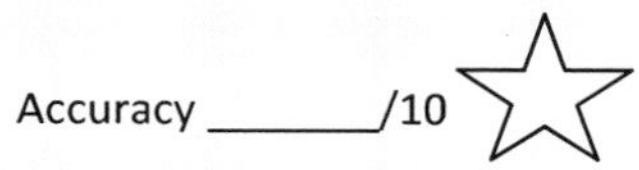

Accuracy ______/10 ☆

DAY 2 – TUESDAY

1	2	3	4	5	6	7	8	9	10
594 406	3095 405	967 137	5156 -4010	1405 99	2199 808	8850 170	3899 1101	6003 97	7555 490

22:12

© SAI Speed Math Academy, USA

Accuracy ______/10 ☆

DAY 3 – WEDNESDAY

1	2	3	4	5	6	7	8	9	10
4070 946	3464 284	4384 620	6470 413	9175 107	6175 328	6811 135	2599 403	7825 -6421	2936 66

22:13

© SAI Speed Math Academy, USA

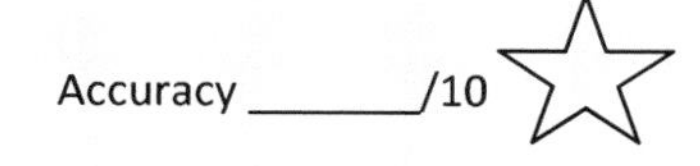

Accuracy ______/10 ☆

DAY 4 – THURSDAY

1	2	3	4	5	6	7	8	9	10
3825 194	670 643	7896 - 582	8496 510	8228 342	6995 08	5110 675	1569 431	6699 302	8065 436

22:14

© SAI Speed Math Academy, USA

DAY 5 – FRIDAY

Accuracy ______/20

1	2	3	4	5	6	7	8	9	10
1996	4657	3802	8898	7655	2006	4999	3907	3339	2335
04	343	198	102	347	996	10	95	1661	2669

22:15

© SAI Speed Math Academy, USA

1	2	3	4	5	6	7	8	9	10
1897	2277	3190	8030	6208	3338	8294	5937	3499	2330
111	412	16	379	85	27	-7030	-4703	50	449
78	41	795	141	707	138	237	-1234	55	721

22:16

© SAI Speed Math Academy, USA

LESSON 17 – DICTATION

DICTATION: Dictation is when teacher or parent calls out a series of numbers and the child listens to the numbers and does the calculation in mind or on the abacus.

DO 6 PROBLEMS A DAY and write answers below.
Students: Try to calculate the dictated problems in mind.
Teachers: Dictate problems from mind math part of this week's work.

1	2	3	4	5	6	7	8	9	10

11	12	13	14	15	16	17	18	19	20

21	22	23	24	25	26	27	28	29	30

ANSWER KEY

WEEK 2

1	2	3	4	5	6	7	8	9	10	
19	20	21	32	40	31	60	49	66	69	2:S:1

1	2	3	4	5	6	7	8	9	10	
195	206	129	220	300	424	319	276	288	620	2:S:2

1	2	3	4	5	6	7	8	9	10	
100	186	112	104	133	142	210	154	81	212	2:S:3

WEEK 3

1	2	3	4	5	6	7	8	9	10	
04	04	33	134	23	34	64	74	145	145	3:S:1

1	2	3	4	5	6	7	8	9	10	
55	64	53	155	64	153	144	254	530	644	3:S:2

1	2	3	4	5	6	7	8	9	10	
94	82	83	245	124	164	145	553	230	257	3:S:3

WEEK 4

1	2	3	4	5	6	7	8	9	10	
10	11	20	35	48	35	58	35	85	80	4:S:1

1	2	3	4	5	6	7	8	9	10	
200	108	190	300	165	290	412	311	275	600	4:S:2

1	2	3	4	5	6	7	8	9	10	
160	266	130	59	129	220	164	256	113	253	4:S:3

WEEK 5

1	2	3	4	5	6	7	8	9	10	
03	12	43	82	41	141	173	240	210	143	5:S:1

1	2	3	4	5	6	7	8	9	10	
114	51	131	145	151	54	164	164	233	553	5:S:2

1	2	3	4	5	6	7	8	9	10	
193	141	103	135	132	133	449	553	533	522	5:S:3

WEEK 6

1	2	3	4	5	6	7	8	9	10	
11	30	35	46	61	62	33	171	178	220	6:S:1

1	2	3	4	5	6	7	8	9	10	
224	227	174	114	131	121	116	176	165	115	6:S:2

1	2	3	4	5	6	7	8	9	10	
108	165	150	280	59	90	111	360	373	425	6:S:3

WEEK 7

1	2	3	4	5	6	7	8	9	10	
20	02	34	42	82	62	220	129	114	209	7:S:1

1	2	3	4	5	6	7	8	9	10	
62	73	56	59	31	122	152	222	210	510	7:S:2

1	2	3	4	5	6	7	8	9	10	
216	232	136	120	227	95	331	201	536	655	7:S:3

WEEK 9

1	2	3	4	5	6	7	8	9	10	
10	15	100	34	25	45	60	65	82	40	9::S:1

1	2	3	4	5	6	7	8	9	10	
101	165	115	180	110	113	121	220	213	120	9::S:2

1	2	3	4	5	6	7	8	9	10	
90	160	160	94	110	250	165	157	130	30	9::S:3

WEEK 10

1	2	3	4	5	6	7	8	9	10	
01	10	21	132	123	131	190	171	124	187	10S:1

1	2	3	4	5	6	7	8	9	10	
13	71	114	21	111	163	124	214	552	534	10S:2

1	2	3	4	5	6	7	8	9	10	
24	44	151	105	112	61	105	251	215	511	10S:3

WEEK 11

1	2	3	4	5	6	7	8	9	10	
14	22	15	45	76	112	130	192	324	542	11::S:1

1	2	3	4	5	6	7	8	9	10	
51	50	55	110	104	107	165	163	306	551	11::S:2

1	2	3	4	5	6	7	8	9	10	
170	120	154	154	105	127	64	310	450	550	11::S:3

WEEK 12

1	2	3	4	5	6	7	8	9	10	
12	11	14	23	100	130	105	160	328	341	12:S:1

1	2	3	4	5	6	7	8	9	10	
61	50	67	53	102	95	154	152	540	410	12:S:2

1	2	3	4	5	6	7	8	9	10	
115	116	123	246	254	24	111	125	105	313	12:S:3

WEEK 13

1	2	3	4	5	6	7	8	9	10	
09	11	65	115	120	70	101	102	125	271	13:S:1

1	2	3	4	5	6	7	8	9	10	
71	52	86	120	105	151	141	95	622	512	13:S:2

1	2	3	4	5	6	7	8	9	10	
162	150	143	124	188	140	120	251	241	453	13:S:3

WEEK 14

1	2	3	4	5	6	7	8	9	10	
08	50	65	81	100	138	00	121	566	421	14:S:1

1	2	3	4	5	6	7	8	9	10	
71	50	30	160	31	200	151	161	550	520	14:S:2

1	2	3	4	5	6	7	8	9	10	
163	221	220	111	122	245	252	121	524	101	14:S:3

WEEK 15

1	2	3	4	5	6	7	8	9	10	
32	40	55	100	105	104	154	140	104	406	15:S:1

1	2	3	4	5	6	7	8	9	10	
50	50	120	10	60	150	120	120	450	510	15:S:2

1	2	3	4	5	6	7	8	9	10	
131	209	121	108	130	104	70	180	340	310	15:S:3

WEEK 16

1	2	3	4	5	6	7	8	9	10	
100	101	102	103	104	105	106	107	108	105	16:S:1

1	2	3	4	5	6	7	8	9	10	
104	103	104	102	103	104	101	102	103	104	16:S:2

1	2	3	4	5	6	7	8	9	10	
500	501	502	503	504	505	506	507	508	503	16:S:3

1	2	3	4	5	6	7	8	9	10	
504	503	504	502	503	504	501	502	503	504	16:S:4

1	2	3	4	5	6	7	8	9	10	
100	110	90	100	101	111	101	91	102	112	16:S:5

1	2	3	4	5	6	7	8	9	10	
92	102	103	113	93	103	104	114	104	94	16:S:6

1	2	3	4	5	6	7	8	9	10	
105	115	95	105	106	116	106	96	107	117	16:S:7

1	2	3	4	5	6	7	8	9	10	
97	107	108	118	108	98	104	114	94	104	16:S:8

1	2	3	4	5	6	7	8	9	10	
104	113	104	93	102	113	94	102	102	104	16:S:9

1	2	3	4	5	6	7	8	9	10	
504	503	504	603	602	413	503	602	502	604	16:S:10

www.abacus-math.com

WEEK 21

1	2	3	4	5	6	7	8	9	10	
1600	1605	1001	1025	1020	1556	1136	1055	1050	1642	21:S:1

1	2	3	4	5	6	7	8	9	10	
1400	1411	1464	1451	1435	1126	1102	1340	1355	1319	21:S:2

1	2	3	4	5	6	7	8	9	10	
5000	5313	5400	5202	5100	5248	5304	5202	5103	4454	21:S:3

1	2	3	4	5	6	7	8	
1647	1502	1504	1190	1537	1883	1446	1302	21:S:4

1	2	3	4	5	6	7	8	
2025	1240	2131	1609	1200	1805	1340	1534	21:S:5

WEEK 21 – PRACTICE WORK

DAY 1 – MONDAY

1	2	3	4	5	6	7	8	
1834	2261	1471	1502	1913	1312	500	1140	21:1

1	2	3	4	5	6	7	8	
3900	1855	4581	2611	9390	5050	8502	5501	21:2

DAY 2 – TUESDAY

1	2	3	4	5	6	7	8	
404	2395	1572	1873	501	1979	2302	1487	21:3

1	2	3	4	5	6	7	8	
5284	8396	4302	5500	5307	4344	8506	5502	21:4

DAY 3 – WEDNESDAY

1	2	3	4	5	6	7	8	
2281	1521	1716	1727	1215	912	1445	1180	21:5

1	2	3	4	5	6	7	8	
5151	5169	5200	6026	5490	4509	5104	8811	21:6

DAY 4 – THURSDAY

1	2	3	4	5	6	7	8	
2400	2154	2802	1602	1051	1460	1272	1507	21:7

1	2	3	4	5	6	7	8	
7819	7247	5401	5504	5115	0000	4204	3151	21:8

DAY 5 – FRIDAY

1	2	3	4	5	6	7	8	
2508	567	1765	2638	3430	2115	2509	2554	21:9

WEEK 21

SEPARATE SHAPES

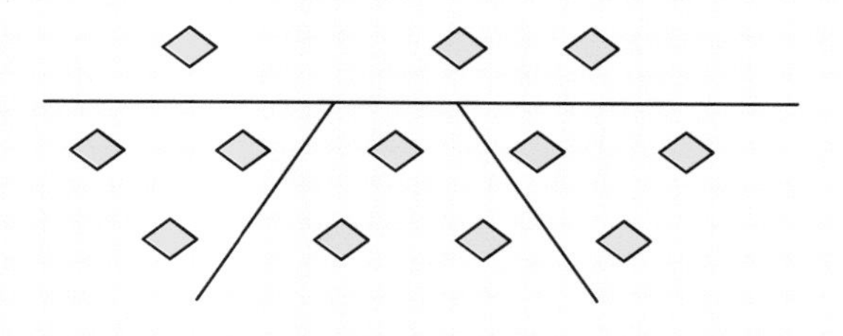

1	2	3	4	5	6	7	8	
6504	9999	3239	5509	2200	5917	5250	5202	21:10

WEEK 21 – MIND MATH PRACTICE WORK

DAY 1 – MONDAY

1	2	3	4	5	6	7	8	9	10	
1597	4073	2901	3554	5066	2980	1101	1355	1527	2909	21:11

DAY 2 – TUESDAY

1	2	3	4	5	6	7	8	9	10	
1100	1280	3020	4240	1461	2700	1528	1379	1101	1521	21:12

DAY 3 – WEDNESDAY

1	2	3	4	5	6	7	8	9	10	
1954	1161	1531	1410	4269	1322	1209	1110	3801	1500	21:13

DAY 4 – THURSDAY

1	2	3	4	5	6	7	8	9	10	
5144	2020	1602	5314	1012	2002	1400	1352	1104	1154	21:14

DAY 5 – FRIDAY

1	2	3	4	5	6	7	8	9	10	
1040	5618	1602	1500	3714	1101	2759	1054	910	2250	21:15

1	2	3	4	5	6	7	8	9	10	
801	1234	1638	1405	5010	1201	702	5455	5555	1182	21:16

WEEK 22

1	2	3	4	5	6	7	8	9	10	
503	1035	1005	1025	1050	1020	1031	1010	1025	1050	22:S:1

1	2	3	4	5	6	7	8	9	10	
1000	1001	1007	1000	1001	1000	1005	1005	1006	1004	22:S:2

1	2	3	4	5	6	7	8	9	10	
2000	5005	5001	5006	5005	5003	5002	5000	5002	4007	22:S:3

1	2	3	4	5	6	7	8	
5004	5004	5002	5001	5000	5003	5000	4014	22:S:4

1	2	3	4	5	6	7	8	
4205	5003	4243	7031	5002	8104	6010	6004	22:S:5

WEEK 22 – PRACTICE WORK

DAY 1 – MONDAY

1	2	3	4	5	6	7	8	
200	502	1000	1031	392	500	2000	554	22:1

1	2	3	4	5	6	7	8	
8003	8003	8470	2232	9004	5000	000	5000	22:2

DAY 2 – TUESDAY

1	2	3	4	5	6	7	8	
400	2010	303	551	00	405	1218	2003	22:3

1	2	3	4	5	6	7	8	
6148	5583	5503	5103	5004	5002	4000	6502	22:4

DAY 3 – WEDNESDAY

1	2	3	4	5	6	7	8	
1595	2091	798	600	01	1301	202	174	22:5

1	2	3	4	5	6	7	8	
5502	00	3504	1001	2920	5450	5552	5000	22:6

DAY 4 – THURSDAY

1	2	3	4	5	6	7	8	
00	402	371	00	1000	612	1502	100	22:7

1	2	3	4	5	6	7	8	
2402	4320	5004	5500	6008	4002	4106	4000	22:8

DAY 5 – FRIDAY

1	2	3	4	5	6	7	8	
505	202	41	554	100	1050	1903	2006	22:9

1	2	3	4	5	6	7	8	
5110	1040	5004	7014	2500	1320	4400	3151	22:10

WEEK 22 – MIND MATH PRACTICE WORK

DAY 1 – MONDAY

1	2	3	4	5	6	7	8	9	10	
501	1000	1040	1000	4040	5001	5006	5000	6000	4006	22:11

DAY 2 – TUESDAY

1	2	3	4	5	6	7	8	9	10	
1000	3500	1104	1146	1504	3007	9020	5000	6100	8045	22:12

DAY 3 – WEDNESDAY

1	2	3	4	5	6	7	8	9	10	
5016	3748	5004	6883	9282	6503	6949	3002	1404	3002	22:13

DAY 4 – THURSDAY

1	2	3	4	5	6	7	8	9	10	
4019	1313	7314	9006	8570	7003	5785	2000	7001	8501	22:14

DAY 5 – FRIDAY

1	2	3	4	5	6	7	8	9	10	
2000	5000	4000	9000	8002	3002	5009	4002	5000	5004	22:15

1	2	3	4	5	6	7	8	9	10	
2086	2730	4001	8550	7000	3503	1501	00	3604	3500	22:16

WORD SEARCH

							H	E	R	D						**L**	
				M				E		**S**	H	E	E	P		A	
				R				S								N	
C				A				E	N	E			L			D	
O	L	A	F	**F**	U	**B**		E		E	K				L		
W								R	L		**H**	C		E			
				B		R	E	**T**	S	O	O	**R**	I				S
		G			M		T		G	O	**H**	**F**		H			S
		I				A									**C**		A
		P			**C**		**L**										R
										E	S	R	O	**H**			**G**

MIRROR IMAGE

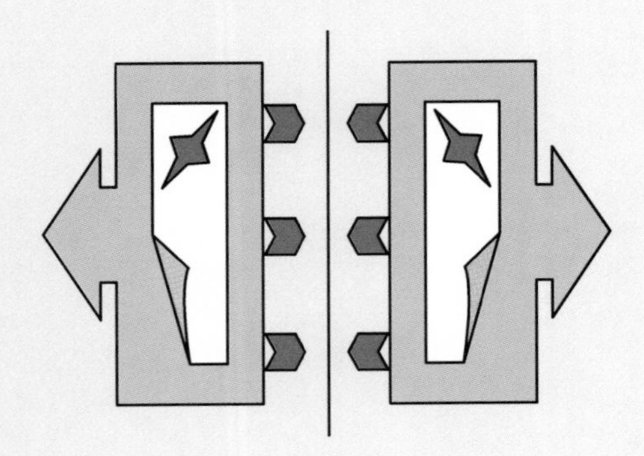

SEPARATE SHAPES

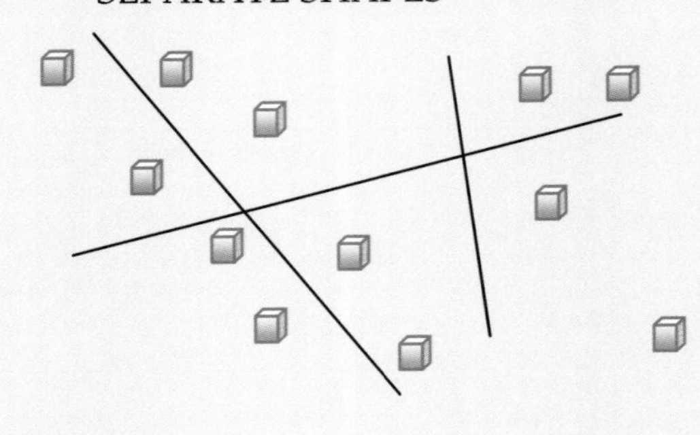

NUMBER BLOCK

				20
5	6	8	1	20
8	6	9	4	27
6	5	7	0	18
5	8	1	9	23
24	25	25	14	27

NUMBER BLOCK

				22
2	6	8	7	23
7	5	1	3	16
9	5	3	8	25
9	8	7	4	28
27	24	19	22	14

SHAPES PUZZLE

= 08 = 05

= 09 = 12

ABOUT SAI SPEED MATH ACADEMY

One subject that is very important for success in this world, along with being able to read and write, is the knowledge of numbers. Math is one subject which requires proficiency from anyone who wants to achieve something in life. A strong foundation and a basic understanding of math is a must to mastering higher levels of math.

We, the family, best friends, and parents of children in elementary school, early on discovered that what our children were learning at school was not enough for them to master the basics of math. Teachers at school, with the resources they had, did the best they could. But, as parents, we had to do more to help them understand the relationship between numbers and basic functions of adding, subtracting, multiplying and dividing. Also, what made us cringe is the fact that our children's attitude towards more complex math was to say, "Oh, we are allowed to use a calculator in class". This did not sit well with us. Even though we did not have a specific system that we followed, each of us could do basic calculations in our minds without looking for a calculator. So, this made us want to do more for our children.

We started to look into the various methods that were available in the marketplace to help our children understand basic math and reduce their dependency on calculators. We came across soroban, a wonderful calculating tool from Japan. Soroban perfectly fits with the base-10 number system used at present and provides a systematic method to follow while calculating in one's mind.

This convinced us and within a short time we were able to work with fluency on the tool. The next step was to introduce it to our children, which we thought was going to be an easy task. It, however, was not. It was next to impossible to find the resources or the curriculum to help us introduce the tool in the correct order. Teaching all the concepts in one sitting and expecting children to apply them to the set of problems we gave them only made them push away the tool in frustration.

However, help comes to those who ask, and to those who are willing to work to achieve their goals. We came across a soroban teacher who helped us by giving us ideas and an outline of how soroban should be introduced. But, we still needed an actual worksheet to give our children to practice on. That is when we decided to come up with practice worksheets of our own design for our kids.

Slowly and steadily, practicing with the worksheets that we developed, our children started to get the idea and loved what they could do with a soroban. Soon we realized that they were better with mind math than we were.

Today, 6 years later, all our kids have completed their soroban training and are reaping the benefits of the hard work that they did over the years.

Now, although very happy, we were humbled at the number of requests we got from parents who wanted to know more about our curriculum. We had no way to share our new knowledge with them.

Now, through the introduction of our instruction book and workbooks, that has changed. We want to share everything we know with all the dedicated parents who are interested in teaching soroban to their children. This is our humble attempt to bring a systematic instruction manual and corresponding workbook to help introduce your children to soroban.

What started as a project to help our kids has grown over the years and we are fortunate to say that a number of children have benefitted learning with the same curriculum that we developed for our children.

Thank you for choosing our system to enhance your children's mathematical skills.

We love working on soroban and hope you do too!

List of SAI Speed Math Academy Publications

LEVEL – 1

Abacus Mind Math Instruction Book Level – 1: Step by Step Guide to Excel at Mind Math with Soroban, a Japanese Abacus
ISBN-13: 978-1941589007

Abacus Mind Math Level – 1 Workbook 1 of 2: Excel at Mind Math with Soroban, a Japanese Abacus

ISBN-13: 978-1941589014

Abacus Mind Math Level – 1 Workbook 2 of 2: Excel at Mind Math with Soroban, a Japanese Abacus

ISBN-13: 978-1941589021

LEVEL – 2

Abacus Mind Math Instruction Book Level – 2: Step by Step Guide to Excel at Mind Math with Soroban, a Japanese Abacus
ISBN-13: 978-1941589038

Abacus Mind Math Level – 2 Workbook 1 of 2: Excel at Mind Math with Soroban, a Japanese Abacus

ISBN-13: 978-1941589045

Abacus Mind Math Level – 2 Workbook 2 of 2: Excel at Mind Math with Soroban, a Japanese Abacus

ISBN-13: 978-1941589052

LEVEL – 3

Abacus Mind Math Instruction Book Level – 3: Step by Step Guide to Excel at Mind Math with Soroban, a Japanese Abacus
ISBN-13: 9781941589069

Abacus Mind Math Level – 3 Workbook 1 of 2: Excel at Mind Math with Soroban, a Japanese Abacus

ISBN-13: 9781941589076

Abacus Mind Math Level – 3 Workbook 2 of 2: Excel at Mind Math with Soroban, a Japanese Abacus

ISBN-13: 9781941589083

Printed in Great Britain
by Amazon